Commercial Electrical Inspector

WB-E2-15

Practice Exam Questions

Mar 2015
COPYRIGHT © 2015
by
Burger, Inc.

ALLRIGHTS RESERVED

2015 Commercial Electrical Inspector Practice Exam Questions

$ISBN_{10}$: 0-9792191-0-8

$ISBN_{13}$: 978-0-9792191-0-8

Printed in USA

IMPORTANT NOTICES AND DISCLAIMERS CONCERNING THIS BOOK

NOTICE AND DISCLAIMER OF LIABILITY CONCERNING THE USE OF THESE DOCUMENTS

Every effort has been made to produce this product to help prepare for the Inspector Exam.

The codes, standards, recommended practices, and guides, on which this document is base is the 2014 National Electrical Code®, as published by the National Fire Protection Association ®.

IBC®, ICC®, and International Building Code® are all registered trademarks and the property of the International Code Council. NFPA and National Electrical Code® is a registered trademark of the National Fire Protection Association®.

Burger, Inc. does not claim to have nor does it have any affiliation with ICC® or NFPA®

Burger, Inc disclaims liability for any personal injury, property or other damages of any nature whatsoever, whether, special, indirect, consequential or compensatory, directly or indirectly resulting from the publication, use of, or reliance on this document. Burger, Inc makes no guaranty or warranty as to the accuracy or completeness of any information published herein.

The sole purpose of this book is to help practice for the ICC® Commercial Electrical Inspector Exam. This Document should not be used as a substitute for the standard Code book or for any other purpose. You must use the NEC® for any code related reference need. By making this document available, Burger, Inc. is not undertaking to render professional or other services for or on behalf of any person or entity. Nor is Burger, Inc undertaking to perform any duty owed by any person or entity to someone else. Anyone using this document should rely on his or her own independent judgment or, as appropriate, seek the advice of a competent professional in determining the exercise of reasonable care in any given circumstances.

Commercial Electrical Inspector Practice Exam Questions
435 Unique Questions

There are 80 questions on the E-2 Exam.

That is equivalent to taking the exam over **5 times!!**

The Result: Passed

All Questions are based on the 2014 National Electrical Code®

by Cliff Burger, ICC Certifications held: B-1, E-1, M-1, P-1, B-2   Other: CompTIA A+, Network+, 30 years in the Construction Industry as a Building Contractor, Master Electrician, Electrical Contractor and Businessman.

This study practice guide contains two sections. Section One contains 415 unique questions covering the subject matter from the International Building. Section Two has the questions set up in 5 complete Timed Exams of 80 questions each, just like the real exam.

Studying the code can help you pass the exam, but because of the extremely large amount of data, exceptions and variations it is impossible to have all that information flawlessly in your mind. That is why the tests are open book. The best inspectors rely on the code book, not their recollection of the code. It is extremely important to be *right* on the safety issues in the code. If you have a basic understanding of terminology you should easily be able to pass the exams by having a fast system down for finding the answers in the book.

The best way to study for an open book test is by doing open book tests. By reading a question and then pursuing the answer your mind is forced to interpret what you read, process that information, and find a solution through recall, and research. These study exam questions will help you by improving your skill of three major things needed to pass the exam:

1. You will increase your speed at finding the answers and you will become an expert on where to find the answers in the code book, this is very important because this is the only way to be correct 100% of the time.

2. You will learn the answers to many of the questions that will be on the test and thereby be able to answer them from **recall** memory, saving you precious time.

3. You will become an expert at managing that time during the test, through test taking experience.

The Test Questions Answers and References are in back of the book.

You can **Pass** all the exams first try with these study questions!

Good Luck!

How to Pass Your Exams First Time

Studying the code can help you pass the exam, but because of the extremely large amount of data, exceptions and variations it is impossible to have all that information flawlessly in your mind. That is why the tests are open book. The best inspectors rely on the code book, not their recollection of the code. It is extremely important to be *right* on the safety issues in the code. If you have a basic understanding of terminology you should easily be able to pass the exams by having a fast system down for finding the answers in the book.

First, do not read the code through intensely paying attention to every detail. There is way too much information to absorb in that way. Instead, first skim through the table of contents to see how the code is arranged. In an open book test it is very important to be able to find the answer and find it fast. Spend 10 minutes and just familiarize yourself with the arrangement of the whole code book. Look though the table of contents. Go to the sections that interest you and find them. Next, go to the back of the book and study the index. Most of the answers you need to find are only seconds away once you find the right topic in the index. You must become an expert at finding what you need in the index. The answers to almost all of the questions on the exam can be found in a matter of a minute or less if you find it first in the index and then in the section that the index points to.

Second, skim through all of the sections in the code paying attention to the titles of each section and subsection lightly skimming the content. This does two things. It further familiarizes you with the code book and how it is arranged and also trains your eyes and mind to be able to go quickly through the content. You will pick up bits and pieces of what you have skimmed and when you see numbers your mind will naturally be inquisitive to see what the numbers apply to.

Let's face it, on an open book test the main thing to overcome is the time factor. All of the answers that you need are right there in that book, you just need to know where and how to find them. If there were no time limit everyone should be able to get 100% correct answers on the test every time.

How To Learn
One thing that they do not teach you in school is *how to learn and how to remember.* Too much emphasis is put on what you are learning and no emphasis is put on a system of learning.

Schools should first teach you how to learn, just like they teach you how to read or do math. Once you have mastered the system of how to learn, you can learn and retain anything.

Most people think that in order to memorize something you need to read it over and over again. While this method will cause memorization over time, it is really an inefficient use of time. In reality memorization is a product of two things, time and recall. If you read a fact or figure and then never make any attempt to bring that information back to your conscious mind, you probably will not remember it very long. Think about how many people you have met only to forget their name. Why because you did not use their name. Remember, if you don't use it, you lose it. Sound familiar, well, it's true.

So I'm learning codes, how do I use them?

Here is how to do it!

Never spend more than about 20 minutes at one time studying through the code book. Why? It is a proven fact that most people concentrate the best for about 20 minutes and then their learning productivity starts to decrease. You will actually learn much more by studying in twenty minute intervals. During the break times away from the code book you will actually keep learning and really learning it if you do this. Try to recall as many of the facts that you have learned during your 20 minute study block as you can. Every time you recall a fact it makes that fact much easier to recall the next time.

The secret is recall over time though not just recall. If you take in any bit of information an within 20 minutes recall that information, then recall it again this time say 3 hours later. You are apt to remember it tomorrow, so if you recall it tomorrow, you are apt to remember it a week from now. Now if you recall it a week from now you will then be able to recall it a month and then a year and the 10 years or for life.

Look at what you did. You read something once, remembered it 6 or 7 times, and now it is imbedded in your memory for life. It actually was imbedded in your mind, the first time you read it. You just needed to then, train your mind how to find it up there in that mass of gray matter.

The first recall is the most important, because if you wait to long your mind my not find the path to the information. Then you have to look it up again and start over. You will learn from experience how long to wait between learning and recalls, 1,2,3,4, 5 ......

So through recall over time you have trained your mind where to find the information that it stores. This simple technique will work for anything you ever need to learn for the rest of your life. The same exact principles can be used on the code book.

Section One in the Book.

Now that you have a good foundation on how to learn and have reviewed the table of contents, the index and skimmed the whole book once, you are ready to start using the practice questions in section one. Use the same twenty minute period method to answer the first practice questions in the workbook. You will probably be able to find from 3 to 6 answers at first in that twenty minute block, and that will keep improving as you practice your skills. Now use the recall technique between each block, reviewing the questions and answers in your mind. You can be watching TV, eating going for a walk, bike riding playing with the kids, visiting with friends, or doing nearly anything during this time, actually diversion and distractions are a good thing when it comes to recall. If you can recall under those conditions, think how easy it would be on a test with no distractions. Who knew learning could be this much fun?

Going in 20 minute blocks using recall go through all of the practice questions. Remember to think about questions and answers that you did in the last 20 minute block. By the time you are done with the main question area in the book, you will be pretty good at finding answers in the code book, and you will have memorized the answers to many of the questions on the exam. Remember, If you can answer them from memory, accurately you save more time for the ones that you have to look up in the book.

The Second Section, Timed Exams

The timed exams will reinforce all of the skills that you are developing and prepare you for the against the clock aspect of the test. The whole time you are building skills in finding answers, memorizing answers through recall and building confidence. By test day you have become a pro at taking the tests and probably know about one third or more of the answers on the test without even looking them up. And, as for the one's you need to look up, you are very good at finding those answers quickly by now.

It's time to start your first 20 minute block, and happy learning.

E-2 Commerical Electrical Inspector Exam Outline
Reference: 2014 National Electrical Code

E2 Commercial Electrical Inspector		80 multiple-choice questions
		Exam fee: $189
		Open book—3-1/2-hour time limit
Content Area	**% of Total**	**Reference**
General Requirements	10%	2014 *National Electrical Code*
Services	17%	
Branch Circuit and Feeder Requirements	19%	
Wiring Methods and Distribution Systems	19%	
General Use Equipment	12%	
Special Occupancies	9%	
Special Equipment	7%	
Special Systems	7%	
Total	100%	

Section 1

435 Practice Questions

1. Nonconductive coatings (such as paint, lacquer, and enamel) on equipment to be grounded shall be removed from threads and other contact surfaces to ensure good electrical _____ or be connected by means of fittings designed so as to make such removal unnecessary.

   A. conduction
   B. voltage
   C. current
   D. continuity

2. Does a pipe organ have any special grounding requirements?

   A. Yes
   B. No

3. Where a portable generator, rated 15 kW or less, is installed using a flanged inlet or other cord- and plug-type connection, a disconnecting means shall _____ where un- grounded conductors serve or pass through a building or structure.

   A. not be required
   B. be required
   C. be readily accessible
   D. provided within 6 ft. of

4. A 3000 sq. ft. store, has 30 ft of show window. There are a total of 80 duplex receptacles. The service is 120/240 V, single phase 3-wire service. Actual connected lighting load is 8500 VA. What is the total calculated receptacle load in VA

   A. 14,400
   B. 10,000
   C. 2,200
   D. 12,200

5. Show windows shall be calculated in accordance with either of the following either the unit load per outlet as required in other provisions of this section or at 200 volt-amperes per _____ ft of show window.

   A. 1
   B. 2
   C. 3
   D. 4

6. General-use dimmer switches can be used to control permanently installed unless listed for the control of other loads and installed accordingly.

   A. paddle fans
   B. fluorescent lights
   C. lighting receptacle outlets
   D. incandescent luminaries

7. Conductor sizes are expressed in American Wire Gage (AWG) or in _____.

   A. sq. inches
   B. standard metric sizes
   C. standard English system sizes
   D. circular mils

8. Grounding electrode conductor(s) shall be installed in one continuous length without a splice or joint except splicing of the wire-type grounding electrode conductor shall be permitted by irreversible compression-type connectors listed as grounding and bonding equipment or by _____.

   A. soldering
   B. silver soldering
   C. the exothermic welding process
   D. arc welding

9. At least _____ entrance of sufficient area shall be provided to give access to and egress from working space about electrical equipment.

   A. 1
   B. 2
   C. 3
   D. 4

10. Continuous Duty

   A. Operation at a substantially constant load for an indefinitely long time.
   B. Operation at a substantially constant load for an 2 or more hours.
   C. Operation at a substantially constant voltage for an indefinitely long time.
   D. Operation at a substantially constant load for the normal operating requirements of the device.

11. All electric pool water heaters shall have the heating elements subdivided into loads not exceeding _____ amperes and protected at not over _____ amperes.

   A. 20, 40
   B. 40, 80
   C. 48, 60
   D. 50, 80

12. A building or other structure served shall be supplied by _____ service. Exceptions ignored

   A. only one
   B. only two
   C. only three
   D. only four

13. The branch circuit serving emergency lighting and power circuits _____ part of a multiwire branch circuit.

   A. shall be
   B. shall not be
   C. can be
   D. is sometimes

14. Bonding jumpers shall be of copper or other corrosion-resistant material, a bonding jumper shall be a _____ or similar suitable conductor.

   A. wire
   B. bus
   C. screw
   D. any of the above

15. What is the adjustment factor percentage for more than three current-carrying conductors in a raceway or cable if there are 21 current-carrying conductors in a 2 inch EMT.

   A. 35
   B. 40
   C. 45
   D. 70

16. Wiring located within the cavity of a fire-rated floor-ceiling or roof-ceiling assembly shall be secured to, or supported by, an independent means of secure support. Where independent support wires are used, they shall be _____. Exceptions ignored.

   A. galvanized steel No. 10 or larger.
   B. steel, copper, brass or aluminum
   C. distinguishable by twists within the bottom 12 inches of the wires
   D. distinguishable by color, tagging, or other effective means from those that are part of the fire-rated design

17. For pools, spas, fountains, and similar locations, lighting systems shall be installed not less than _____ horizontally from the nearest edge of the water, unless permitted by Article 680.

   A. 9 m (30 ft)
   B. 6 m (20 ft)
   C. 12 m (40 ft)
   D. 3 m (10 ft)

18. A means shall be provided in each metal box for the connection of an equipment grounding conductor. The means shall be permitted to be a _____ or equivalent.

   A. tapped hole
   B. toggle bolt
   C. cable clamp
   D. strap

19. The grounding electrode conductor shall be _____. or the items as permitted in 250.68(C).

   A. copper, aluminum, or copper-clad aluminum
   B. solid or stranded, and insulated
   C. stranded, insulated, covered, or bare
   D. solid or stranded, and bare

20. In rigid metal conduit or intermediate metal conduit, what is the minimum burial depth for wiring under streets, highways, roads, alleys, driveways, and parking lots.

   A. 6
   B. 12
   C. 18
   D. 24

21. Where a flexible cord is used to supply a room air conditioner, the length of such cord shall not exceed _____ ft for a nominal, 120-volt rating _____ ft for a nominal, 208- or 240-volt rating.

   A. 8, 10
   B. 6, 4
   C. 10, 6
   D. 6, 3

22. For each yoke or strap containing one or more devices or equipment, a _____ volume allowance shall be made for each yoke or strap based on the largest conductor connected to a device(s) or equipment supported by that yoke or strap.

   A. single
   B. double
   C. zero
   D. multiple

23. _____ of conductors on buildings, structures, or poles shall be as provided for services in 230.50.

   A. Proper spacing
   B. Insulating values
   C. Mechanical protection
   D. Minimum height clearances

24. The copper sheath of mineral-insulated, metal-sheathed cable Type MI may be used as the equipment grounding conductor.

   A. This is true
   B. This is false

25. Fuses and circuit breakers shall be permitted to be connected in parallel where they are _____ and listed as a unit. Individual fuses, circuit breakers, or combinations thereof shall not otherwise be connected in parallel.

   A. rated equally for amperage and time delay
   B. matched pairs
   C. factory assembled in parallel
   D. of the same type and rating

26. The minimum cover requirements for underground wiring 0 to 600 Volts, in Type UF cable under streets, highways, roads, alleys, driveways, and parking lots is _____ inches.

   A. 6
   B. 12
   C. 18
   D. 24

27. A 3000 sq. ft. store, has 30 ft of show window. There are a total of 80 duplex receptacles. The service is 120/240 V, single phase 3-wire service. Actual connected lighting load is 8500 VA. What is the total calculated general lighting load in VA.

   A. 6000
   B. 9000
   C. 12000
   D. 16200

28. High-impedance grounded neutral systems in which a grounding impedance _____ ground-fault current to a low value shall be permitted for 3-phase ac systems of 480 volts to 1000 volts where certain conditions are met.

   A. usually a resistor
   B. of not less greater that 10 ohms
   C. of not greater than 10 mA.
   D. of not greater than 240 Volts.

29. The maximum current, in amperes, that a conductor can carry continuously under the conditions of use without exceeding its temperature rating.

   A. Ampacity
   B. Voltage
   C. Power Factor
   D. Capacitance

30. Where a branch circuit supplies continuous loads or any combination of continuous and noncontinuous loads, the rating of the overcurrent device shall not be less than the noncontinuous load plus _____ percent of the continuous load.

    A. 80
    B. 100
    C. 115
    D. 125

31. All switchboards, switch-gear, panel-boards, distribution boards, and motor control centers shall be located in _____ spaces and protected from damage.

    A. interior
    B. exterior
    C. dedicated
    D. dry

32. Lighting track may be installed in the following locations.

    A. In storage battery rooms
    B. In wet or damp locations
    C. Where likely to be subjected to physical damage
    D. Where concealed
    E. none of the above

33. Paper insulated conductors typically have what outer covering:

    A. Copper or alloy steel
    B. None
    C. Lead sheath
    D. Glass braid

34. In dwelling units, where a _____ is installed in an island or peninsular countertop and the width of the countertop behind the range, counter-mounted cooking unit, or sink is less than 300 mm (12 in.), the _____ is considered to divide the countertop space into two separate countertop spaces.

    A. wall outlets
    B. wall switches
    C. pictures
    D. range, counter-mounted cooking unit, or sink

35. A 18 AWG fixture wire shall be permitted to be tapped to the branch circuit conductor of a In a 20 amp branch circuit provided it is _____ ft. or less in length.

    A. 6
    B. 10
    C. 25
    D. 50

36. XHH insulation has what which characteristic trade name.

    A. Moisture-resistant thermoset
    B. Modified ethylene tetrafluoro-ethylene
    C. Extended polytetra-fluoro-ethylene
    D. Thermoset

37. The distance between a cable or conductor entry and its exit from the box shall not be less than _____ times the outside diameter, over sheath, of that cable or conductor. Exceptions ignored.

    A. 24
    B. 30
    C. 36
    D. 40

38. The dwelling has a floor area of 1500 sq. ft. exclusive of an unfinished cellar not adaptable for future use, unfinished attic, and open porches. It has two 20-A small appliance circuits, one 20-A laundry circuit, two 4-kW wall-mounted ovens, one 5.1-kW counter-mounted cooking unit, a 4.5-kW water heater, a 1.2-kW dishwasher, a 5-kW combination clothes washer and dryer, six 7-A, 230-V room air-conditioning units, and a 1.5-kW, 230 V permanently installed bathroom space heater. Assume wall-mounted ovens, counter-mounted cooking unit, water heater, dishwasher, and combination clothes washer and dryer kW ratings equivalent to kVA. What is the total calculated load for the service in VA and the total required minimum service rating, assuming that the two 4-kVA wall-mounted ovens are supplied by one branch circuit, the 5.1-kVA counter-mounted cooking unit by a separate circuit.

    A. Calculated Service load 29,200 VA - Minimum Service Rating 122A
    B. Calculated Service load 15,200 VA - Minimum Service Rating 100A
    C. Calculated Service load 39,200 VA - Minimum Service Rating 150A
    D. Calculated Service load 5,200 VA - Minimum Service Rating 60A

39. A grounding rod electrode shall be permitted to be buried in a trench that is at least _____ in. deep, and the upper end of the electrode shall be flush with or below ground level unless the aboveground end and the grounding electrode conductor attachment are protected against physical damage as specified in 250.10.

    A. 60
    B. 48
    C. 36
    D. 30

40. Fixed or stationary equipment other than an underwater luminaire (lighting fixture) for a permanently installed pool shall be permitted to be connected with a _____.

    A. overhead service drop
    B. 6 ft. long flexible cord
    C. flexible cord to facilitate the removal or disconnection for maintenance or repair
    D. none of the above

41. Continuous Load is a load where the maximum current is expected to continue for ____ hours or more.

   A. 1
   B. 2
   C. 3
   D. 4

42. Underground service conductors shall be insulated for _____.

   A. workers safety
   B. expose to the elements
   C. the applied voltage
   D. identification purposes

43. Handles or levers of circuit breakers, and similar parts that may move suddenly in such a way that persons in the vicinity are likely to be injured by being struck by them, shall _____.

   A. not be used
   B. be made non-accessible
   C. locked in the open position
   D. be guarded or isolated

44. For circuits supplying loads consisting of motor-operated utilization equipment that is fastened in place and has a motor larger than ____ hp in combination with other loads, the total calculated load shall be based on _____ percent of the largest motor load plus the sum of the other loads.

   A. 1/8, 125
   B. 1/4, 110
   C. 1/2, 150
   D. 3/4, 80

45. In or under airport runways, including adjacent areas where trespassing prohibited what is the minimum burial depth for direct burial cables or conductors.

    A. 6
    B. 12
    C. 18
    D. 24

46. Switches shall not be installed within tub or shower spaces _____.

    A. unless they are grounded, bonded and GFI protected
    B. unless they have a watertight cover
    C. unless installed as part of a listed tub or shower assembly.
    D. under any circumstances

47. What is the maximum ampere rating of a insulated single copper conductor in air with a conductor temperature of 90 C. in an ambient temperature of 40 C , the conductor size is No. 4 and the voltage is 2500 Volts.

    A. 100
    B. 110
    C. 115
    D. 145

48. For permanently connected appliances rated at not over 300 volt-amperes or 1/8hp, the branch-circuit overcurrent device shall _____.

    A. be within sight of the appliance
    B. not be permitted to serve as the disconnecting means
    C. be permitted to serve as the disconnecting means
    D. none of the above

49. Flexible Metallic Tubing (FMT) can be used in which of the following places or conditions.

   A. In hoistways
   B. Under ground for direct earth burial, or embedded in poured concrete or aggregate
   C. For system voltages of 1000 volts maximum
   D. In lengths over 100 ft

50. In dwelling units the lighting demand factor of the first 3000 VA or less is _____ Percent.

   A. 50
   B. 80
   C. 100
   D. 125

51. Where a change occurs in the size of the ungrounded conductor, a similar change shall be permitted to be made in the size of the _____ conductor.

   A. grounded
   B. grounding
   C. bonding
   D. neutral

52. The dwelling has a floor area of 1500 sq. ft. exclusive of an unfinished cellar not adaptable for future use, unfinished attic, and open porches. It has two 20-A small appliance circuits, one 20-A laundry circuit, two 4-kW wall-mounted ovens, one 5.1-kW counter-mounted cooking unit, a 4.5-kW water heater, a 1.2-kW dishwasher, a 5-kW combination clothes washer and dryer, six 7-A, 230-V room air-conditioning units, and a 1.5-kW, 230 V permanently installed bathroom space heater. Assume wall-mounted ovens, counter-mounted cooking unit, water heater, dishwasher, and combination clothes washer and dryer kW ratings equivalent to kVA. Assuming the dwelling is feed by a 120/240-V, 3-wire, single-phase service what is the total calculated load in VA and the minimum required service rating required.

   A. 19,200 VA, 100 A
   B. 25,700 VA, 110 A
   C. 29,200 VA, 122 A
   D. 39,200 VA, 175 A

53. Vegetation such as trees shall _____ for support of overhead conductor spans.

   A. be used only temporarily
   B. be branches minimum 4 inches in diameter
   C. not be used
   D. be pruned clean and used

54. The electric vehicle supply equipment shall be located for direct electrical coupling of the EV con- nector (conductive or inductive) to the electric vehicle. Unless specifically listed and marked for the location, the coupling means of the electric vehicle supply equipment shall be stored or located at a height of not less than _____ above the floor level for indoor locations and _____ above the grade level for outdoor locations.

   A. 24 in , 36 in
   B. 18 in , 24 in
   C. 36 in , 48 in
   D. any height is acceptable

55. Openings through which conductors enter a box shall be _____.

   A. sealed
   B. clamped shut
   C. open for ventilation
   D. closed in an approved manner

56. Overhead Service, where not in excess of 1000 volts, nominal, shall follow minimum clearances measured from _____.

   A. the top of the mast
   B. final grade
   C. the guy wire anchor
   D. the foundation wall

57. Electrical equipment shall be firmly secured to the surface on which it is mounted, and wooden plugs driven into holes in masonry, concrete, plaster, or similar materials shall _____.

    A. Be oversized to provide a tight fit.
    B. be used to insulate the equipment from the mounting surface.
    C. a minimum of 3/4" diameter
    D. not be used

58. A receptacle installed outdoors in a location protected from the weather or in other damp locations shall have an enclosure for the receptacle that is weatherproof when the receptacle is _____.

    A. being used by a cord plugged into it
    B. open
    C. covered
    D. energized

59. What is the adjustment factor percentage for more than three current-carrying conductors in a raceway or cable if there are 4 current-carrying conductors in a 3/4 inch EMT.

    A. 45
    B. 50
    C. 70
    D. 80

60. All storage or instantaneous-type water heaters shall be equipped with a temperature-limiting means in addition to its control thermostat to disconnect all _____.

    A. connected conductors
    B. ungrounded conductors
    C. water supply systems
    D. none of the above

61. The means of coupling to the electric vehicle shall be either conductive or inductive. At- tachment plugs, electric vehicle connectors, and electric vehicle inlets shall be listed or labeled for the purpose. The overall usable length shall not exceed _____ unless equipped with a cable management system that is part of the listed electric vehicle supply equipment.

    A. 8 ft
    B. 15 ft
    C. 25 ft
    D. 50 ft

62. It shall be permissible to apply a demand factor of 75 percent to the nameplate rating load of _____ or more appliances fastened in place, other than electric ranges, clothes dryers, space-heating equipment, or air-conditioning equipment, that are served by the same feeder or service in a one-family, two-family, or multifamily dwelling.

    A. one
    B. two
    C. three
    D. four

63. THHN insulation has what which characteristic trade name.

    A. Heat-resistant thermoplastic
    B. Moisture-resistant thermoplastic
    C. Moisture- and heat-resistant thermoplastic
    D. Moisture, heat, and oil resistant thermoplastic

64. Switching devices shall be located at least _____ ft horizontally from the inside walls of a pool unless separated from a pool by a solid fence, wall, or other permanent barrier. Alternatively, a switch that is listed as being acceptable for use within this distance shall be permitted.

    A. 5
    B. 10
    C. 15
    D. 20

65. Snap switches, including dimmer and similar control switches, shall be connected to an equipment grounding conductor and shall provide a means to connect metal faceplates to the equipment grounding conductor, whether or not a metal faceplate is installed. Exception ignored.

    A. This is True to the Code
    B. This is False to the Code

66. For dwelling units, attached garages, and detached garages with electric power, at least one _____ shall be installed to provide illumination on the exterior side of outdoor entrances or exits with grade level access.

    A. Flood light
    B. Motion detector
    C. Security camera
    D. wall switch controlled lighting outlet

67. An LB is an example of a _____.

    A. Conduit Body
    B. Fitting
    C. Connector
    D. Junction Box

68. A wall space in a dwelling unit for purposes of outlet spacing shall be considered any unbroken any space _____ ft or more in width including space measured around corners and unbroken along the floor line by doorways, fireplaces, and similar openings.

    A. 1
    B. 2
    C. 3
    D. 4

69. A warehouse storage area uses mercury vapor lighting. What is the minimum lighting load in the warehouse area if it has 40,500 sq. ft. for storage.

   A. 10125 VA
   B. 7875 VA
   C. 6625 VA
   D. 5165 VA

70. Bends in Types NM, NMC, and NMS cable shall be so made that the cable will not be damaged. The radius of the curve of the inner edge of any bend during or after installation shall not be less than _____ times the diameter of the cable.

   A. ten
   B. twenty
   C. five
   D. thirty six

71. The common point on a wye-connection in a polyphase system or midpoint on a single-phase, 3-wire system, or midpoint of a single-phase portion of a 3-phase delta system, or a midpoint of a 3-wire, direct-current system.

   A. Plenum
   B. Switching Device
   C. Service Point
   D. Neutral Point.

72. Unbroken lengths of busway shall be permitted to be extended through _____.

   A. dry walls
   B. hoistways
   C. hazardous locations
   D. damp locations

73. A complete lighting unit consisting of a light source such as a lamp or lamps, together with the parts designed to position the light source and connect it to the power supply. It may also include parts to protect the light source or the ballast or to distribute the light.

    A. Luminaire
    B. Lampholder
    C. Outline Lighting
    D. Lamp

74. The minimum cover requirements for underground wiring 0 to 600 Volts, in rigid metal conduit under a building is _____ inches.

    A. 0
    B. 6
    C. 12
    D. 18

75. In a Cellular Metal Floor Raceway the combined cross-sectional area of all conductors or cables shall not exceed _____ percent of the interior cross-sectional area of the cell or header.

    A. 25
    B. 40
    C. 60
    D. 75

76. Ground Fault

    A. Connected (connecting) to ground or to a conductive body that extends the ground connection.
    B. Connected to ground without inserting any resistor or impedance device.
    C. An unintentional, electrically conducting connection between an ungrounded conductor of an electrical circuit and the normally non-current-carrying conductors, metallic enclosures, metallic raceways, metallic equipment, or earth.
    D. The circuit conductors between the final overcurrent device protecting the circuit and the outlet(s).

77. A multi-family dwelling has 40 dwelling units. Meters are in two banks of 20 each with individual feeders to each dwelling unit. One-half of the dwelling units are equipped with electric ranges not exceeding 12 kW each. Assume range kW rating equivalent to kVA rating in accordance with 220.55. Other half of ranges are gas ranges. Area of each dwelling unit is 840 sq. ft. Laundry facilities on premises are available to all tenants. Add no circuit to individual dwelling unit. What is the general lighting load for each unit, in VA.

   A. 1520
   B. 2050
   C. 2520
   D. 3630

78. Type UF cable may be used as follows:

   A. In storage battery rooms
   B. As service-entrance cable
   C. Between a dwelling and separate garage unit
   D. Where exposed to direct rays of the sun, unless identified as sunlight resistant

79. On overhead service-drop individual conductors shall be insulated or covered, however, the _____ conductor of a multiconductor cable shall be permitted to be bare.

   A. grounding
   B. isolated hot leg
   C. ungrounded
   D. grounded

80. For circuits supplying lighting units that have ballasts, transformers, or autotransformers, the calculated load shall be based on _____

    A. total watts of the lamps
    B. efficiency ratio of the luminaire
    C. the total ampere ratings of such units and not on the total watts of the lamps
    D. number of lamps per unit x the wattage of each lamp

81. The purpose of the National Electrical Code is the _____ of persons and property from hazards arising from the use of electricity.

    A. Economical protection
    B. Protection
    C. Adequate Safety
    D. Practical Safeguarding.

82. The branch-circuit conductors supplying one or more units of a information technology equipment shall have an ampacity not less than _____ percent of the total connected load.

    A. 80
    B. 100
    C. 125
    D. 150

83. Conductors shall be considered outside of a building or other structure where they are installed under not less than _____ in. of concrete beneath a building or other structure or where the are installed within a building or other structure in a raceway that is encased in concrete or brick not less than _____ in. thick

    A. 3, 3
    B. 4, 4
    C. 6, 6
    D. 2, 2

84. A 3000 sq. ft. store, has 30 ft of show window. There are a total of 80 duplex receptacles. The service is 120/240 V, single phase 3-wire service. Actual connected lighting load is 8500 VA. What is the total calculated show window lighting load in VA.

    A. 6000
    B. 9000
    C. 5000
    D. 7000

85. What is the approximate diameter in inches of No. 2 RHW wire.

    A. .412
    B. .375
    C. .275
    D. .125

86. A 125-volt, single-phase, 15- or 20-ampere-rated receptacle outlet shall be installed at an accessible location for the servicing of heating, air-conditioning, and refrigeration equipment, and the receptacle shall be located on the same level and within _____ of the heating, air-conditioning, and refrigeration equipment.

    A. 25 ft
    B. 15 ft
    C. 10 ft
    D. 6 ft

87. The required branch circuit copper conductor size for taps of a 30 amp rated circuit is No. _____.

    A. 14
    B. 12
    C. 10
    D. 8

88. Wiring located within the cavity of a fire-rated floor-ceiling or roof-ceiling assembly shall not be secured to, or supported by _____. Exceptions ignored.

    A. metal cross members
    B. the ceiling assembly, including the ceiling support wires
    C. joists, beams, or columns
    D. cantilevered floor joists

89. Lighting busway and trolley busway shall not be installed less than _____ ft above the floor or working platform unless provided with a cover identified for the purpose.

    A. 8
    B. 10
    C. 12
    D. 14

90. In kitchens, pantries, breakfast rooms, dining rooms, and similar areas of dwelling units, _____ small-appliance branch circuits shall serve all wall and floor receptacle outlets and all countertop outlets and receptacle outlets for refrigeration equipment. Exceptions ignored.

    A. one
    B. two
    C. two or more 20-ampere
    D. at least three

91. Which of the following shall not be used as grounding electrodes:

    A. Metal underground gas piping system
    B. Aluminum electrodes
    C. Concrete rebar
    D. A and B

92. Raceways, cable assemblies, boxes, cabinets, and fittings shall be securely fastened in place. Support wires that do not provide secure support shall not be _____.

    A. permitted
    B. allowed to be counted as part of the structural system of fastening
    C. permitted as the sole support
    D. copper or aluminum

93. In dwelling units, in every kitchen, family room, dining room, living room, parlor, library, den, sunroom, bedroom, recreation room, or similar room or area of dwelling units, receptacle outlets shall be installed so that no point measured horizontally along the floor line in any wall space is more than _____ ft from a receptacle outlet.

    A. 2
    B. 12
    C. 3
    D. 6

94. Liquidtight Flexible Nonmetallic Conduit (LFNC) is permitted to be used _____. Exceptions Ignored.

    A. Where protection of the contained conductors is required from vapors, liquids, or solids
    B. In lengths longer than 100 ft,
    C. Where the operating voltage of the contained conductors is in excess of 600 volts, nominal,
    D. In any conditions of extreme cold or heat.

95. What is the ampacity rating of a two conductor copper No. 8 Type SCE flexible cord rated a 60 C. with an ambient temperature o 30 C.

    A. 60
    B. 55
    C. 48
    D. 65

96. What is the maximum percent of FMC conduit total interior area that can be filled if more that 2 conductors are used in the tube.

   A. 53
   B. 31
   C. 40
   D. 60

97. Receptacles rated 20 amperes or less and designed for the direct connection of aluminum conductors shall be marked _____.

   A. CO/ALR
   B. Aluminum
   C. Any Metal
   D. Bi-Metal

98. A unit load of not less than _____ per square foot is required for general lighting in Church occupancies.

   A. 1 VA
   B. 2 VA
   C. 3 VA
   D. 4 VA

99. A ground ring encircling the building or structure, in direct contact with the earth, consisting of at least 20 ft of bare copper conductor not smaller than _____ AWG shall be permitted to be used as a grounding electrode.

   A. 2/0
   B. 1
   C. 2
   D. 4

100. A device that provides a means for connecting intersystem bonding conductors for communications systems to the grounding electrode system.

   A. Grounding Electrode Conductor
   B. Intersystem Bonding Termination.
   C. Grounding Electrode
   D. Explosionproof Apparatus

101. What is the maximum allowable distance between trade size 1 RMC supports in feet.

   A. 10
   B. 12
   C. 14
   D. 16

102. What is the maximum number of conductors or fixture wires that can be used in electrical metallic tubing for No. 14 Type THHW conductors in trade size 27 tubing.

   A. 6
   B. 10
   C. 16
   D. 28

103. The NEC (National Electrical Code) is not intended as a design specification or an instruction manual for _____ .

   A. electrical engineers
   B. untrained persons
   C. power linemen
   D. none of the above

104. The connection of a grounding electrode conductor or bonding jumper to a grounding electrode shall be accessible except, an encased or buried connection in a _____, driven, or buried grounding electrode shall not be required to be accessible.

   A. rigid metallic tube
   B. concrete-encased
   C. flexible metallic tube
   D. electrical metallic tube

105. FMC is not permitted in _____.

   A. In storage battery rooms
   B. Concealed spaces
   C. Exposed locations
   D. Interior walls

106. The overhead conductors between the utility electric supply system and the service point.

    A. Service Entrance
    B. Service Drop
    C. Service Lateral
    D. Power Entrance

107. It shall be permissible to calculate the load of a feeder or service that supplies three or more dwelling units of a multifamily dwelling in accordance with Table 220.84 instead of Part III of this article if all the following condition(s) is (are) met.

    A. Each dwelling unit is equipped with electric resistance heating systems.
    B. No dwelling unit is supplied by more than one feeder
    C. Each dwelling unit is equipped with electric cooking equipment.
    D. B and C

108. No parts of cord-connected luminaries (fixtures), chain-, cable-, or cord-suspended-luminaries (fixtures), lighting track, pendants, or ceiling-suspended (paddle) fans shall be located within a zone measured _____ ft horizontally and _____ ft vertically from the top of the bathtub rim or shower stall threshold.

    A. 3, 8
    B. 5, 8
    C. 5, 10
    D. 8, 8

109. The branch-circuit rating for an appliance that is a continuous load, other than a motor-operated appliance, shall not be less than _____ percent of the marked rating, or not less than 100 percent of the marked rating if the branch-circuit device and its assembly are listed for continuous loading at 100 percent of its rating.

    A. 100
    B. 125
    C. 150
    D. 175

110. Where a feeder supplies continuous loads or any combination of continuous and noncontinuous loads, the rating of the overcurrent device shall not be less than the noncontinuous load plus _____ percent of the continuous load. Exceptions ignored.

   A. 80
   B. 100
   C. 125
   D. 150

111. Circuit conductors shall be permitted to be installed in raceways; in cable trays; as metal-clad cable, as bare wire, cable Thpe MC: as bare wire, cable, and busbars; or as Type MV cables or conductors as provided in 300.37, 300.39, 300.40, and 300.50. Bare live conductors shall _____.

   A. not be allowed
   B. conform with 490.24.
   C. be installed in PVC RNC.
   D. be marked as live.

112. Outlet boxes that do not enclose devices or utilization equipment shall have a minimum internal depth of _____.

   A. 3/4
   B. 1
   C. 1/2
   D. 15/16

113. All wet niche swimming pool luminaries shall be removable from the water for relamping or normal maintenance. Luminaries shall be installed in such a manner that personnel can reach the luminaire for relamping, maintenance, or inspection while on the deck or equivalently _____.

   A. wet location
   B. dry location
   C. accessible location
   D. none of the above

114. An uninsulated equipment grounding conductor shall be permitted, but, if individually covered, the covering shall have a continuous outer finish that is _____. Exceptions ignored.

    A. green
    B. green with yellow strips
    C. yellow with green strips
    D. A or B

115. A device, group of devices, or other means whereby the conductors of a circuit can be disconnected from their source of supply.

    A. Switch
    B. Disconnecting Means
    C. Closed Circuit Switch
    D. Relay Switch

116. The greatest root-mean-square (rms) (effective) difference of potential between any two conductors of the circuit concerned.

    A. Amperage
    B. Wattage
    C. Voltage
    D. Power Factor

117. Lampholders of the screw-shell type shall be installed for use as lampholders only. Where supplied by a circuit having a grounded conductor, the grounded conductor shall be connected to the _____.

    A. center terminal
    B. green terminal
    C. grounding terminal
    D. screw shell

118. Where a feeder overcurrent device is not readily accessible, branch-circuit overcurrent devices shall be installed on the load side, shall be mounted in a readily accessible location, and shall be of _____ ampere rating than the feeder overcurrent device.

    A. a lower
    B. a higher
    C. the same
    D. 150 percent of the

119. The building or structure disconnecting means shall plainly indicate whether it is in the _____ position.

    A. hot or cold
    B. normal
    C. correct
    D. open or closed

120. Electrical Nonmetallic Tubing, ENT may be used _____. Exceptions ignored.

    A. Where the voltage is over 600 volts
    B. Where subject to physical damage
    C. In a two story building
    D. For direct earth burial

121. ENT shall not be used for the _____ of luminaries and other equipment.

    A. connection
    B. wiring method
    C. support
    D. conductors

122. _____ meeting the requirements of this article shall be used around impaired connections, such as reducing washers or oversized, concentric, or eccentric knockouts. Standard locknuts or bushings shall not be the only means for the bonding required by this section but shall be permitted to be installed to make a mechanical connection of the raceway(s).

    A. Locknuts
    B. Copper bushings
    C. Compression fittings
    D. Bonding jumpers

123. A grounding electrode system consisting of a ground ring shall be buried at a depth below the earth's surface of not less than _____ in.

   A. 18
   B. 24
   C. 30
   D. 60

124. Underground service conductors shall have sufficient ampacity to carry the current for the load as calculated and the minimum size of the conductors shall not be smaller than _____ aluminum or copper-clad aluminum.

   A. 6 AWG copper or 6 AWG
   B. 8 AWG copper or 6 AWG
   C. 4 AWG copper or 6 AWG
   D. 8 AWG copper or 8 AWG

125. The load for household electric clothes dryers in a dwelling unit(s) shall be either _____ watts or the nameplate rating, whichever is larger, for each dryer served.

   A. 4000
   B. 5000
   C. 7500
   D. 10,000

126. All 120-volt, single phase, 15- and 20-ampere branch circuits supplying outlets or devices installed in dwelling unit kitchens, family rooms, dining rooms, living rooms, parlors, libraries, dens, bedrooms, sunrooms, recreation rooms, closets, hallways, or similar rooms or areas shall be protected by a listed _____ installed to provide protection of the branch circuit.

   A. ground fault interupter
   B. fuse
   C. arc-fault circuit interrupter, combination-type,
   D. bonded jumper

127. For installations to supply only limited loads of a single branch circuit, the branch circuit disconnecting means shall have a rating of not less than _____ amperes.

   A. 10
   B. 15
   C. 20
   D. 30

128. In dwelling units, no small-appliance branch circuit shall _____.

   A. serve more than two kitchens
   B. serve more than one dining room
   C. serve a more than one kitchen
   D. serve a dining room and a kitchen

129. Means shall be provided for disconnecting all _____ conductors that supply or pass through the building or structure.

   A. grounded
   B. ungrounded
   C. grounding
   D. bonding

130. A reliable conductor to ensure the required electrical conductivity between metal parts required to be electrically connected.

   A. Bonding Jumper
   B. Connection
   C. Soldering
   D. Welding

131. The width of the working space in front of the electric equipment shall be the width of the equipment or _____ in., whichever is greater and the work space shall permit at least a _____ degree opening of equipment doors or hinged panels.

   A. 90, 30
   B. 45, 60
   C. 30, 90
   D. 60, 45

132. A snap switch shall not be grouped or ganged in enclosures with other snap switches, receptacles, or similar devices, unless they are arranged so that the voltage between adjacent devices does not exceed _____ volts, or unless they are installed in enclosures equipped with identified, securely installed barriers between adjacent devices.

    A. 120
    B. 240
    C. 300
    D. 600

133. In circuits for control of irrigation and landscape lighting limited to not more than 30 volts and installed with type UF cable under minimum of 4-in. thick concrete exterior slab with no vehicular traffic and the slab extending not less than 6 in. beyond the underground installation what is the minimum required burial depth.

    A. 6
    B. 12
    C. 18
    D. 24

134. In a service installation for a dwelling what is the maximum service feeder rating for 4/0 aluminum service conductors in amperes.

    A. 150
    B. 175
    C. 200
    D. 225

135. A complete lighting unit consisting of a light source such as a lamp or lamps, together with the parts designed to position the light source and connect it to the power supply. It may also include parts to protect the light source or the ballast or to distribute the light.

    A. Light Fixture
    B. Incandescent Lamp
    C. Street Lamp
    D. Luminaire

136. A applied to circuit breakers, a qualifying term indicating that there is purposely introduced a delay in the tripping action of the circuit breaker, which delay decreases as the magnitude of the current increases.

    A. Delayed Trip
    B. Slow Blow
    C. Time Protected
    D. Inverse Time

137. The grounding electrode shall be _____ the grounding electrode conductor connection to the system.

    A. within 10 feet of
    B. within the same general area but no case more than 30' from
    C. as near as practicable to, and preferably in the same area as,
    D. directly adjacent to

138. Circuits that supply electric motor driven fire pumps shall be supervised from _____ disconnection.

    A. inadvertent
    B. overload
    C. water load
    D. all of the above

139. Electrodes of bare or conductively coated iron or steel plates must be at least _____ in. thick and at least _____ sq. feet in contact with exposed earth to be used as a grounding electrode.

    A. 0.06, 3
    B. 0.06, 2
    C. .25, 2
    D. 0.06, 4

140. Continuity of the grounding path or the bonding connection to interior piping shall not rely on _____.

    A. water meters or filtering devices and similar equipment
    B. grounding jumpers
    C. patch conductors and connectors
    D. pressure connectors

141. In walls constructed of wood or other combustible material, cabinets shall be flush with the finished surface _____.

    A. or project therefrom
    B. or recessed not more than 1/4 inch
    C. or set off of the surface 1/4 inch.
    D. placed on a metal plate

142. Admitting close approach; not guarded by locked doors, elevation, or other effective means.

    A. approve
    B. accessible
    C. workable
    D. open

143. Electrical equipment and wiring and other electrically conductive material likely to become energized shall be installed in a manner that creates a low-impedance circuit facilitating the operation of the overcurrent device or ground detector for high-impedance grounded systems. It shall be capable of safely carrying the maximum ground-fault current likely to be imposed on it from any point on the wiring system where a ground fault may occur to the electrical supply source.

    A. GFI
    B. Bonding
    C. Effective Ground-Fault Current Path.
    D. grounded conductor

144. The minimum number of branch circuits shall be determined from the total _____ load and the size or rating of the circuits used.

    A. calculated
    B. area of
    C. amperage
    D. voltage

145. The equipment grounding conductor can be the casing enclosing the circuit conductors and may be in the form of any of the following except :_____

   A. RMC
   B. RNC
   C. IMT
   D. EMT

146. Where space limitations prevent wiring from being routed a distance 5 ft or more from a pool, such wiring shall be permitted where installed in complete raceway systems of _____.

   A. rigid metal conduit
   B. intermediate metal conduit
   C. or a nonmetallic raceway system
   D. none of the above
   E. all of the above

147. Where cable is used, each cable shall be _____ to the cabinet, cutout box, or meter socket enclosure. Exceptions ignored

   A. secured
   B. clamped
   C. bonded
   D. affixed

148. For receptacles in damp or wet locations, an installation suitable for wet locations shall also be considered suitable for _____ locations.

   A. underwater
   B. damp
   C. exposed
   D. corrosive vapor

149. Metric designator 41 is the same as trade size _____ inch conduit.

   A. 1/2
   B. 1
   C. 1 1/2
   D. 2

150. _____ shall be provided to give safe access to the working space around electric equipment installed on platforms, balconies, or mezzanine floors or in attic or roof rooms or spaces.

    A. Elevators
    B. Permanent ladders or stairways
    C. Temporary ladders
    D. Collapsible stairways

151. A cable and components used to ensure survivability of critical circuits for a specified time under fire conditions

    A. Asbestos Shielded Cable System
    B. Gemini Duplicated Cable System
    C. Refractory Shielded Cable Sytem
    D. Fire-Resistive Cable System.

152. A multi-family dwelling has 40 dwelling units. Meters are in two banks of 20 each with individual feeders to each dwelling unit. One-half of the dwelling units are equipped with electric ranges not exceeding 12 kW each. Assume range kW rating equivalent to kVA rating in accordance with 220.55. Other half of ranges are gas ranges. Area of each dwelling unit is 840 sq. ft. Laundry facilities on premises are available to all tenants. Add no circuit to individual dwelling unit. Could you use 2 15 A. circuits to handle the general lighting load?

    A. Yes
    B. No

153. Where necessary to prevent tampering, an automatic overcurrent device that protects service conductors supplying only a specific load, such as a water heater, shall be permitted to be _____ where located so as to be accessible.

    A. locked or sealed
    B. openly accessible only
    C. hidden from view
    D. not labeled

154. Circuit breakers shall be marked with their ampere rating in a manner that will be durable and visible after installation, the marking shall be permitted to be made visible by _____.

    A. tripping the breaker
    B. removal of a trim or cover
    C. removing the breaker
    D. markings on the cover

155. An apparatus designed to control and organize unused lengths of output cable to the electric vehicle.

    A. Energy conservation control system (Electric Vehicle Supply Equipment)
    B. Energy Management control board system (Electric Vehicle Supply Equipment)
    C. Green vehicle CO2 footprint optimization system (Electric Vehicle Supply Equipment)
    D. Cable Management System (Electric Vehicle Supply Equipment)

156. Signs and outline lighting systems shall be marked with the manufacturer's name, trademark, or other means of identification, and _____.

    A. input voltage
    B. current rating
    C. date of manufacturing
    D. A & B
    E. none of the above

157. Where outdoor lampholders are attached as pendants, the connections to the circuit wires shall be _____.

    A. staggered
    B. aligned
    C. taped
    D. sleeved

158. Overhead spans of open conductors and open multiconductor cables shall have a vertical clearance of not less than _____ ft above the roof surface. The vertical clearance above the roof level shall be maintained for a distance not less than ft in all directions from the edge of the roof. Exceptions ignored.

    A. 8, 3
    B. 12, 6
    C. 12, 4
    D. 8, 10

159. _____ connectors shall not be concealed.

    A. rigid
    B. XLT
    C. angle
    D. 45 degree offset

160. A switch intended for use in general distribution and branch circuits. It is rated in amperes, and it is capable of interrupting its rated current at its rated voltage.

    A. Isolation Switch
    B. General-Use Switch
    C. Toggle Switch
    D. Breaker Switch

161. How much free space in cubic inches is required in a octagon box that has 6 - No. 12 wires entering and terminating in it. There is also 3 - grounding conductors terminating in it. There is also a fixture stud and three entry clamps inside of the box.

    A. 18.5
    B. 20
    C. 20.25
    D. 21.75

162. Bonding Conductor or Jumper

A. Performs a function without the necessity of human intervention
B. A generic term for a group of nonflammable synthetic chlorinated hydrocarbons used as electrical insulating media.
C. The circuit conductors between the final overcurrent device protecting the circuit and the outlet(s).
D. A reliable conductor to ensure the required electrical conductivity between metal parts required to be electrically connected.

163. Constructed, protected, or treated so as to prevent rain from interfering with the successful operation of the apparatus under specified test conditions.

A. Raintight
B. Waterproof
C. Rainproof
D. Watertight

164. With respect to Solar Photovoltaic (PV) Systems equipment, overcurrent device ratings shall be not less than _____ of the maximum currents calculated in 690.8(A)., Exceptioned ignored.

A. 100 percent
B. 125 percent
C. 150 percent
D. 200 percent

165. Where more than one ground rod is used for grounding the electrodes, each electrode of one grounding system (including that used for strike termination devices) shall not be less than _____ ft from any other electrode of another grounding system.

A. 4
B. 6
C. 8
D. 10

166. An outdoor branch circuit has conductors of 600 V and is strung between poles over a water area not suitable for boating. The conductors are a messenger wire supported twisted cable unit and if 15 ft above the water. You have inspected the installation and come to the following conclusion.

    A. The installation meets Code requirements.
    B. The installation is in violation of Code requirements because the height requirement is not met.
    C. The installation is in violation of Code requirements because the voltage requirement is not met.
    D. none of the above

167. Ground-Fault Circuit Interrupter (GFCI)

    A. A system or circuit conductor that is intentionally grounded.
    B. An unintentional, electrically conducting connection between an ungrounded conductor of an electrical circuit and the normally non-current-carrying conductors, metallic enclosures, metallic raceways, metallic equipment, or earth.
    C. All circuit conductors between the service equipment, the source of a separately derived system, or other power supply source and the final branch-circuit overcurrent device.
    D. A device intended for the protection of personnel that functions to deenergize a circuit or portion thereof within an established period of time when a current to ground exceeds the values established for a Class A device.

168. Conductors in raceways shall be _____ between outlets, boxes, devices, and so forth. Special exceptions ignored.

    A. pulled
    B. installed
    C. continuous
    D. in sections

169. A raceway consisting of grounded metal enclosure containing factory-mounted, bare or insulated conductors, which are usually copper or aluminum bars, rods, or tubes.

    A. Raceway
    B. Collector
    C. Busway
    D. Grounding Bus

170. Enclosures for overcurrent devices shall be mounted in a _____ position unless that is shown to be impracticable.

    A. perpendicular
    B. upright
    C. horizontal
    D. vertical

171. A single receptacle installed on an individual branch circuit shall have an ampere rating of _____ that of the branch circuit.

    A. not less than 100% of
    B. not less than 50% of
    C. not less than 80% of
    D. not less than 125% of

172. Type MV cable shall not be used _____, unless identified for the use.

    A. in wet or dry locations
    B. in raceways
    C. where exposed to direct sunlight
    D. in messenger-supported wiring

173. If required by the authority having jurisdiction, a _____ shall be provided prior to the installation of the feeders.

    A. feeder conductor meter test
    B. diagram showing feeder details
    C. sample calculation
    D. flexible mirror scope inspection device

174. What is the maximum allowable ampacity for No. 12 copper conductors type THHN, 167 F rated insulation, installed in an ambient temperature of 30 C. with 27 current carrying conductors in the same raceway.

    A. 25
    B. 15
    C. 11.25
    D. 18.75

175. A hand-operable circuit breaker equipped with a lever or handle, or a power-operated circuit breaker capable of being opened by hand in the event of a power failure, shall be permitted to serve as a switch if it has _____.

    A. a plastic insulated handle
    B. the required number of poles
    C. not been mounted in a panelboard
    D. has been mounted in a panelboard

176. Where conductors are run in parallel in multiple raceways or cables, the equipment grounding conductors, where used, shall be run _____.

    A. in one of the conductor raceways or cables
    B. in parallel in each raceway or cable
    C. a separate raceway or cable
    D. any of the above

177. Unless required elsewhere in this Code, equipment grounding conductors shall be permitted to be bare, covered, or _____. Exceptions ignored.

    A. painted green
    B. energized
    C. insulated
    D. fused

178. Each conductor that originates outside the box and terminates or is spliced within the box shall be counted once, and each conductor that passes through the box without splice or termination _____.

    A. does not have to be counted
    B. shall be counted twice
    C. shall be counted once
    D. shall be counted once for all such conductors as a group

179. Where the number of current-carrying conductors in a raceway or cable exceeds _____, or where single conductors or multiconductor cables are installed without maintaining spacing for a continuous length than _____ in. and are not installed in raceways, the allowable ampacity of each conductor shall be reduced by a derating adjustment factor. Exceptions ignored.

   A. three, 24
   B. six, 48
   C. twelve, 24
   D. fifteen, 48

180. In a service installation for a dwelling what is the maximum service feeder rating for No. 4 copper service conductors in amperes.

   A. 100
   B. 125
   C. 200
   D. 225

181. What is the minimum bending radius in inches for 3/4" EMT using a Full Shoe bender.

   A. 3.5
   B. 4.5
   C. 5.5
   D. 7.25

182. In swimming pool installations, where none of the required bonded parts is in direct connection with the pool water, the pool water shall be in direct contact with an approved corrosion-resistant conductive surface that exposes not less than _____ of surface area to the pool water at all times. The conductive surface shall be located where it is not exposed to physical damage or dislodgement during usual pool ac- tivities, and

   A. 9 sq. in.
   B. 81 sq. in.
   C. 100 sq. in.
   D. 300 sq. in.

183. Each building or structure disconnect shall _____ disconnect all ungrounded supply conductors it controls and shall have a fault-closing rating not less than the maximum available short-circuit current available at its supply terminals. Exceptions ignored.

    A. automatically
    B. remotely
    C. directly
    D. simultaneously

184. Fuses and circuit breakers shall be _____ so that persons will not be burned or otherwise injured by their operation.

    A. located or shielded
    B. insulated
    C. non-accessible
    D. non of the above

185. Electrical installations in hollow spaces, vertical shafts, and ventilation or air-handling ducts shall be made so that the possible spread of fire or products of combustion will not be substantially increased. Openings around electrical penetrations into or through fire-resistant-rated walls, partitions, floors, or ceilings shall be _____ using approved methods to maintain the fire resistance rating.

    A. blocked
    B. sealed
    C. caulked or taped
    D. firestopped

186. Normally non-current-carrying conductive materials enclosing electrical conductors or equipment, or forming part of such equipment, shall be connected together and to the electrical supply source in a manner that establishes an effective ground-fault current path.

    A. bonding of electrical equipment.
    B. grounded conductor
    C. ungrounded conductor
    D. ground fault

187. Where practicable, rod, pipe, and plate electrodes shall be embedded below _____. Rod, pipe, and plate electrodes shall be free from nonconductive coatings such as paint or enamel.

    A. the average frost line
    B. footings of the structure
    C. surface of the grade line
    D. permanent moisture level

188. A premises wiring system supplied by a grounded ac service shall have a _____ connected to the grounded service conductor, at each service. Exception ignored.

    A. lightening arrester
    B. circuit breaker
    C. grounding electrode conductor
    D. bonding jumper

189. A concrete-encased electrode shall consist of at least 6.0 m (20 ft) of either (1) or (2): (1) One or more bare or zinc galvanized or other electrically conductive coated steel reinforcing bars or rods of not less than 13 mm (12 in.) in diameter, installed in one continuous _____ foot length, or if in multiple pieces connected together by the usual steel tie wires, exothermic welding, welding, or other effective means to create a 6.0 m (20 ft) or greater length; or (2) Bare copper conductor not smaller than 4 AWG Metallic components shall be encased by at least _____ inches of concrete and shall be located horizontally within that portion of a concrete foundation or footing that is in direct contact with the earth or within vertical foundations or structural components or members that are in direct contact with the earth. If multiple concrete-encased electrodes are present at a building or structure, it shall be permissible to bond only one into the grounding electrode system.

    A. 2,16
    B. 3, 20
    C. 4, 20
    D. 20, 2

190. Parts of electric equipment that in ordinary operation produce arcs, sparks, flames, or molten metal shall be enclosed or separated and isolated from all _____.

    A. other electrical components
    B. possible contact with people
    C. combustible material
    D. metal parts

191. Service conductors installed as open conductors or multiconductor cable without an overall outer jacket shall have a clearance of not less than _____ ft from windows that are designed to be opened, doors, porches, balconies, ladders, stairs, fire escapes, or similar locations. Exceptions ignored.

    A. 3
    B. 5
    C. 8
    D. 10

192. A building or structure electrical disconnecting means can be _____.

    A. located where ever practicle
    B. protected from public tampering by locks or disguises
    C. electrically operated by a readily accessible, remote-control device in a separate building or structure.
    D. all of the above

193. Where an ac system is connected to a grounding electrode in or at a building or structure, _____ electrode(s) shall be used to ground conductor enclosures and equipment in or on that building or structure.

    A. separate
    B. independent
    C. non-bonded
    D. the same

194. When calculating conductor fill in a box, each loop or coil of, unbroken conductor not less than twice the minimum length required for free conductors shall be counted _____.

   A. as zero
   B. once
   C. as three
   D. twice

195. A raceway of circular cross section made of helically wound, formed, interlocked metal strip.

   A. Electrical Metallic Tubing (EMT)
   B. Rigid Metal Conduit (RMC)
   C. Flexible Metal Conduit (FMC)
   D. Rigid Non-metallic Conduit (RNC)

196. An enclosure that is designed for either surface mounting or flush mounting and is provided with a frame, mat, or trim in which a swinging door or doors are or can be hung.

   A. Panel Board
   B. Controller
   C. Cabinet
   D. Cutout Box

197. A dwelling has a floor area of 1500 sq. ft, with unfinished cellar not adaptable for future use, unfinished attic, and open porches. Appliances are a 12-kW range and a 5.5-kW, 240-V dryer. Assume range and dryer kW ratings equivalent to kVA ratings in accordance with 220.54 and 220.55. What is the minimum number of branch general light circuits required.

   A. Three 15-A, 2-wire or two 20-A, 2-wire circuits
   B. Five 15-A, 2-wire or three 20-A, 2-wire circuits
   C. Two 15-A, 2-wire and two 20-A, 2-wire circuits
   D. Four 15-A, 2-wire or Three 20-A, 2-wire circuits

198. Small Appliance Circuit Load. In each dwelling unit, the load shall be calculated at _____ for each 2-wire mall-appliance branch circuit as covered by 210.11(C)(1). Where the load is subdivided through two or more feeders, the calculated load for each shall include not less than _____ for each 2-wire smallappliance branch circuit.

    A. 500 volt-amperes, 1500 volt-amperes
    B. 1000 volt-amperes, 1000 volt-amperes
    C. 1500 volt-amperes, 1500 volt-amperes
    D. 2000 volt-amperes, 1000 volt-amperes

199. The point of connection between the facilities of the serving utility and the premises wiring.

    A. Meter Base
    B. Meter Socket
    C. Power Entrance
    D. Service Point

200. The required branch circuit copper conductor size for circuit wires of a 40 amp rated circuit is No. _____.

    A. 12
    B. 10
    C. 8
    D. 6

201. What is the ampacity rating of No. 1 (AWG) copper-clad aluminum conductors at 90 C. in multiconductor cables with not more than three insulated conductors, rated 0 through 2000 Volts, in free air based on ambient air temperature of 40°C for type MC cables

    A. 84
    B. 108
    C. 120
    D. 126

202. _____ shall not be used where conduits or connectors requiring the use of locknuts or bushings are to be connected to the side of the box.

    A. Handy boxes
    B. Octagon boxes
    C. Square boxes
    D. Round boxes

203. For overhead conductors near a swimming pool for Insulated cables, 0-750 Volts to ground, supported on and cabled together with an effectively grounded bare messenger or effectively grounded neutral conductor, what is the clearance in any direction to the observation stand, tower, or diving platform.

    A. 22.5 feet
    B. 25 feet
    C. 14.5 feet
    D. 17 feet

204. A metal underground water pipe in direct contact with the earth for _____ ft or more, including any metal well casing effectively bonded to the pipe, and electrically continuous, or made electrically continuous by bonding around insulating joints or insulating pipe, to the points of connection of the grounding electrode conductor and the bonding conductor(s) or jumper(s), if installed.

    A. 8
    B. 10
    C. 12
    D. 15

205. Overhead conductors for festoon lighting shall not be smaller than 12 AWG unless the conductors are supported by messenger wires. In all spans exceeding _____ ft, the conductors shall be supported by messenger wire.

    A. 40
    B. 50
    C. 60
    D. 100

206. IMC larger than metric designator trade size _____ shall not be used.

   A. 2
   B. 3
   C. 4
   D. 5

207. An outlet on a 20 amp rated circuit must be rated at _____ amps.

   A. 15
   B. 20
   C. 30
   D. Either A or B

208. All _____ of the same circuit and, where used, the grounded conductor and all equipment grounding conductors and bonding conductors shall be contained within the same raceway, auxiliary gutter, cable tray, cablebus assembly, trench, cable, or cord. Exceptions excluded.

   A. control conductors
   B. relay conductors
   C. conductors
   D. communication

209. The grounding electrode conductor shall be of _____. or the items as permitted in 250.68(C).

   A. copper
   B. aluminum
   C. copper-clad aluminum
   D. any of the above

210. Where a raceway enters a building or structure from an underground distribution system, it shall be _____ in accordance with 300.5(G). Spare or unused raceways shall also be _____.

   A. labeled
   B. sealed
   C. listed
   D. insulated

211. Cable tray systems use to support service-entrance conductors shall contain only _____ conductors. Exception ignored.

    A. service-entrance
    B. Type MC cable
    C. branch feeder
    D. insulated

212. Where a building or structure has any combination of feeders, branch circuits, or services _____, a permanent plaque or directory shall be installed at each feeder and branch-circuit disconnect location that denotes all other services, feeders, or branch circuits supplying that building or structure or passing through that building or structure and the area served by each.

    A. passing through or supplying it
    B. passing through it
    C. supplying it
    D. all of the above makes this statement true

213. Overcurrent protection shall be provided in each ungrounded circuit conductor and shall be located _____. Exceptions ignored.

    A. at the point where the conductors receive their supply
    B. at each end of a conductor
    C. at a convent service point in the circuit
    D. on the load end of the circuit

214. Where a conduit enters a box, fitting, or other enclosure, _____ shall be provided to protect the wire from abrasion unless the box, fitting, or enclosure design provides equivalent protection.

    A. a bushing or adapter
    B. a insulated collar
    C. pipe nipple
    D. end clamp

215. In dwelling units, where two or more branch circuits supply devices or equipment on the same yoke or mounting strap, a means to _____ the ungrounded conductors supplying those devices shall be provided at the point at which the branch circuits originate.

    A. connect
    B. simultaneously disconnect
    C. disconnect
    D. energize

216. The identification of terminals to which a grounded conductor is to be connected shall be substantially _____ in color and the identification of other terminals shall be of a readily distinguishable different color.

    A. green
    B. white
    C. gold
    D. black

217. Where the assembly, including the overcurrent devices protecting the feeder(s), is listed for operation at _____ percent of its rating, the allowable ampacity of the feeder conductors shall be permitted to be not less than the sum of the continuous load plus the noncontinuous load.

    A. 80
    B. 100
    C. 125
    D. 150

218. Installations underground or in concrete slabs or masonry in direct contact with the earth; in locations subject to saturation with water or other liquids, such as vehicle washing areas; and in unprotected locations exposed to weather.

    A. Wet Location
    B. Dry Location
    C. Damp Location
    D. High Humidity Location

219. Metal or nonmetallic raceways, cable armors, and cable sheaths shall be _____ between cabinets, boxes, fit-tings, or other enclosures or outlets.

   A. continuous
   B. anchored
   C. supported
   D. horizontal

220. Overhead spans of open conductors and open multiconductor cables of not over 600 volts, nominal, shall have a clearance of not less than _____ ft over public streets, alleys, roads, parking areas subject to truck traffic, driveways on other than residential property, and other land traversed by vehicles, such as cultivated, grazing, forest, and orchard.

   A. 10
   B. 12
   C. 15
   D. 18

221. Liquidtight Flexible Metal Conduit (LFMC) is permitted to be used _____.

   A. Where total bends exceeds 360 degrees.
   B. For direct burial where listed and marked for the purpose
   C. Where subject to physical damage
   D. Where any combination of ambient and conductor temperature produces an operating temperature in excess of that for which the material is approved

222. The metal frame of the building or structure, where _____ ft or more of a single structural metal member in direct contact with the earth or encased in concrete that is in direct contact with the earth shall be permitted to be used as a grounding electrode.

   A. 5
   B. 8
   C. 10
   D. 12

223. Where conductors carrying alternating current are installed in ferrous metal enclosures or ferrous metal raceways, they shall be arranged so as to avoid heating the surrounding ferrous metal by induction by _____ . Exceptions ignored.

    A. installing conductor spacers
    B. coiling like phase conductors
    C. grouping all phase conductors together
    D. none of the above

224. Expansion fittings and telescoping sections of metal raceways shall be made electrically continuous by _____ or other means.

    A. application of oxidation inhibitors
    B. cleaning bright before assembly
    C. equipment bonding jumpers
    D. the use of bare grounding conductors

225. Where rock bottom is encountered in installing a grounding rod electrode the electrode shall be driven at an oblique angle not to exceed _____ degrees from the vertical or, where rock bottom is encountered at an angle up to 45 degrees.

    A. 30
    B. 45
    C. 60
    D. 75

226. Appliance receptacle outlets installed in a dwelling unit for specific appliances, such as laundry equipment, shall be installed within _____ ft of the intended location of the appliance.

    A. 3
    B. 4
    C. 6
    D. 8

227. Two-wire dc circuits and ac circuits of two or more _____ conductors shall be permitted to be tapped from the _____ conductors of circuits that have a grounded neutral conductor.

   A. ungrounded
   B. grounding
   C. grounded
   D. hot leg

228. A multifamily dwelling contains 11 units. What is the demand factor to be used when calculating the service load.

   A. 50
   B. 45
   C. 42
   D. 37

229. What is the maximum allowable ampacity for No. 6 aluminum conductors type THHN, 194 F rated insulation, installed in an ambient temperature of 30 C. There are 3 current carrying conductors in the cable.

   A. 50
   B. 75
   C. 60
   D. 55

230. Except as elsewhere required or permitted by the Code, live parts of electrical equipment operating at _____ volts or more shall be guarded against accidental contact by approved enclosures or by any of the following other means.

   A. 35
   B. 50
   C. 80
   D. 100

231. The connection between two or more portions of the equipment grounding conductor.

    A. Aluminum Wire
    B. Branch Circuit
    C. Copper Wire
    D. Equipment Bonding Jumper

232. Service disconnecting means shall not be installed in _____.

    A. hallways
    B. utility rooms
    C. bathrooms
    D. basements

233. The disconnecting means shall be installed _____ of the building or structure served or where the conductors pass through the building or structure and, the disconnecting means shall be at a readily accessible location nearest the point of entrance of the conductors. Exceptions ignored.

    A. inside
    B. outside
    C. inside and outside
    D. either inside or outside

234. For residential branch circuits rated 120 volts or less with GFIC protection and maximum overcurrent protection of 20 amperes, What is the minimum burial depth In inches in a trench below 2-in. thick concrete or equivalent.

    A. 6
    B. 12
    C. 18
    D. 24

235. With respect to Solar Photovoltaic (PV) Systems equipment, required ground fault protection devices or systems shall:

    A. Be capable of detecting a ground fault in the PV array dc current-carrying conductors and components, in- cluding any intentionally grounded conductors,
    B. Interrupt the flow of fault current
    C. Provide an indication of the fault
    D. Be listed for providing PV ground-fault protection
    E. All of the above

236. All wiring from the controllers to fire pump motors shall be in rigid metal conduit, intermediate metal conduit, electrical metallic tubing, liquidtight flexible metal conduit, or liquidtight flexible nonmetallic conduit Type LFNC-B, listed Type MC cable with an impervious covering, or Type MI cable. Electrical connections at motor terminal boxes shall be made with a listed means of connection such as _____.

    A. twist-on type
    B. insulation-piercing type
    C. soldered wire connectors
    D. none of the above

237. Externally Operable

    A. Equipment enclosed in a case that is capable of withstanding an explosion of a specified gas or vapor that may occur within it and of preventing the ignition of a specified gas or vapor surrounding the enclosure by sparks, flashes, or explosion of the gas or vapor within, and that operates at such an external temperature that a surrounding flammable atmosphere will not be ignited thereby.
    B. Capable of being operated without exposing the operator to contact with live parts.
    C. Capable of being inadvertently touched or approached nearer than a safe distance by a person. It is applied to parts that are not suitably guarded, isolated, or insulated.
    D. Incapable of being operated without exposing the operator to contact with live parts.

238. Raceways shall be provided with expansion fittings where necessary to _____ for thermal expansion and contraction.

   A. provide enhancement
   B. provide restriction
   C. capacitance
   D. compensate

239. A dwelling has a floor area of 1500 sq. ft, with unfinished cellar not adaptable for future use, unfinished attic, and open porches. Appliances are a 12-kW range and a 5.5-kW, 240-V dryer. Assume range and dryer kW ratings equivalent to kVA ratings in accordance with 220.54 and 220.55. What is the total net calculated load for the dwelling in VA units.

   A. 15,600
   B. 30,500
   C. 18,600
   D. 22,600

240. With respect to Solar Photovoltaic (PV) Systems equipment, a device installed in the PV source circuit or PV output circuit that can provide an output dc voltage and current at a higher or lower value than the input dc voltage and current

   A. AC-to-DC Converter.
   B. DC-to-AC Converter.
   C. DC-to-DC Converter.
   D. DC-to-DC Inverter.

241. Devices such as pressure terminal or pressure splicing connectors and soldering lugs shall be identified for the _____ and shall be properly installed and used.

   A. material of the conductor
   B. location of installation
   C. size of conductors
   D. proper spacing

242. Where the opening to an outlet, junction, or switch point is less than _____ in. in any dimension, each conductor shall be long enough to extend at least _____ in. outside the opening. Exception ignored.

    A. 8, 3
    B. 6, 4
    C. 4, 3
    D. 6, 6

243. In general receptacles shall be not less than _____ ft from the inside walls of a pool.

    A. 5
    B. 6
    C. 15
    D. 25

244. Branch circuits that supply neon tubing installations shall not be rated in excess of _____ amperes.

    A. 15
    B. 20
    C. 30
    D. 40

245. What is the minimum bending space at a terminal required for a 4/0 copper conductor with 3 wires per terminal.

    A. 5.5 inches
    B. 8.5 inches
    C. 9 inches
    D. 10 inches

246. Splices or taps shall be permitted within gutters where they are accessible by means of removable covers or doors. The conductors, including splices and taps, shall not fill the gutter to more than _____ percent of its area.

    A. 25
    B. 50
    C. 75
    D. 90

247. This Code does not covers the installation of electrical conductors, equipment, and raceways; signaling and communications conductors, equipment, and raceways; and optical fiber cables and raceways for _____.

    A. Installations of conductors and equipment that connect to the supply of electricity.
    B. Yards, lots, parking lots, carnivals, and industrial substations.
    C. Installations of communications equipment under the exclusive control of communications utilities located outdoors or in building spaces used exclusively for such installations.
    D. Public and private premises, including buildings, structures, mobile homes, recreational vehicles, and floating buildings.

248. An electric power production system that is operating in parallel with and capable of delivering energy to an electric primary source supply system.

    A. Motor Control Center
    B. Multioutlet Assembly
    C. Interactive System
    D. Supplementary Overcurrent Protective Device

249. The complete electrical system shall be performance tested when _____.

    A. before occupancy
    B. first installed on-site
    C. at 3 separate stages during construction.
    D. on a monthly basis

250. Service cables, where subject to physical damage, shall be protected by any of the following except.

    A. Rigid metal conduit
    B. Schedule 80 rigid PVC conduit
    C. Electrical metallic tubing
    D. Intermediate metal conduit

E. all are acceptable

251. Laundry Circuit Load. A load of not less than 1500 volt-amperes shall be included for each 2-wire laundry branch circuit installed, this load shall be permitted to be included with the _____ load and subjected to the demand factors provided in Table 220.42.

    A. overall appliance
    B. utility circuit
    C. heating
    D. general lighting

252. A receptacle outlet shall be installed at each wall countertop space that is _____ or wider. Receptacle outlets shall be installed so that no point along the wall line is more than _____ measured horizontally from a receptacle outlet in that space. Exceptions ignored.

    A. 24 in., 48 in.
    B. 18 in., 36 in.
    C. 12 in., 24 in.
    D. 2 ft., 4 ft.

253. The equipment grounding conductor run with or enclosing the circuit conductors may be _____ conductor.

    A. copper, aluminum, or copper-clad aluminum
    B. copper
    C. aluminum
    D. copper-clad aluminum

254. In kitchens where the receptacles are installed to serve the countertop surfaces outlets within _____ shall be GFI protected.

    A. 3 ft. of the sink
    B. 6 ft. of the sink
    C. 24 ft. of the sink
    D. all these outlets require GFI protection

255. Conductors of branch circuits supplying more than one receptacle for _____ loads shall have an ampacity of not less than the rating of the branch circuit.

    A. permanent
    B. most
    C. cord-and-plug-connected portable
    D. washing machine

256. High Density Polyethylene (HDPE) Conduit a nonmetallic raceway of circular cross section, with associated couplings, connectors, and fittings for the installation of electrical conductors is not allowed to be installed _____.

    A. In discrete lengths or in continuous lengths from a reel
    B. In cinder fill
    C. In locations subject to severe corrosive influences as covered in 300.6 and where subject to chemicals for which the conduit is listed
    D. Within a building

257. Electrical Nonmetallic Tubing (ENT) is not permitted to be used in any building not exceeding _____ floors above grade, unless a fire sprinkler system(s) is installed in accordance with NFPA 13-2002, Standard for the Installation of Sprinkler Systems, on all floors.

    A. one
    B. two
    C. three
    D. five

258. Disconnecting means shall be _____. The provisions for locking shall remain in place with or without the lock installed.

    A. locked from public tampering
    B. lockable in accordance with 110.25.
    C. locked in the closed position
    D. A and C

259. When using Electrical Metallic Tubing: Type EMT wiring systems, when equipment grounding is required, _____

　　A. an equipment ground screw connected to an EMT pipe clamp can be used to ground the equipment.
　　B. EMT will provide sufficient grounding means.
　　C. a separate equipment grounding conductor shall be installed in the conduit.
　　D. any of the above

260. Where a building or structure has any combination of feeders, branch circuits, or services passing through it or supplying it, a permanent plaque or directory shall be installed at each _____ denoting all other services, feeders, or branch circuits supplying that building or structure or passing through that building or structure and the area served by each. Exceptions ignored.

　　A. conductor terminal
　　B. wet location
　　C. feeder and branch-circuit disconnect location
　　D. control panel

261. _____ covers shall be installed on all boxes, fittings, and similar enclosures to prevent accidental contact with energized parts or physical damage to parts or insulation.

　　A. Suitable
　　B. Metal
　　C. Metal or Plastic
　　D. Protective

262. What is Type HFF wiring used for.

　　A. Heat, Fire and Flame resistant
　　B. underground
　　C. Fixture wiring
　　D. hydrogen atmosphere applications

263. Other than for motor overload protection, no overcurrent device shall be connected in series with any conductor that is intentionally grounded, the overcurrent device opens all conductors of the circuit, including the grounded conductor, and is designed so that no pole can operate _____.

   A. simultaneously
   B. above the open limit
   C. until the circuit breaker for each is closed
   D. independently

264. A dwelling has a floor area of 1500 sq. ft, with unfinished cellar not adaptable for future use, unfinished attic, and open porches. Appliances are a 12-kW range and a 5.5-kW, 240-V dryer. Assume range and dryer kW ratings equivalent to kVA ratings in accordance with 220.54 and 220.55. Assuming the dwelling is feed by a 120/240-V, 3-wire, single-phase service, What is the minimum required service size in amperes.

   A. 60
   B. 100
   C. 150
   D. 200

265. Electrical equipment that depends on the natural circulation of air and _____ for cooling of exposed surfaces shall be installed so that room airflow over such surfaces is not prevented by walls or by adjacent installed equipment.

   A. independent fans
   B. heat sinks
   C. convection principles
   D. direct contact of cooling fins

266. Metal raceways, cable trays, cable armor, cable sheath, enclosures, frames, fittings, and other metal noncurrent-carrying parts that are to serve as equipment grounding conductors, with or without the use of supplementary equipment grounding conductors, shall be effectively bonded where necessary to ensure _____ and the capacity to conduct safely any fault current likely to be imposed on them.

    A. ground stability
    B. electrical continuity
    C. electrical insulation
    D. electrical inductance

267. The entrance to all buildings, vaults, rooms, or enclosures containing exposed live parts or exposed conductors operating at over 600 volts, nominal, shall be kept locked unless such entrances are under _____.

    A. the legal occupancy limit of the building
    B. protection of a circuit ground fault device
    C. amperage limit of 1200 amperes
    D. the observation of a qualified person at all times

268. In Dwelling units, permanently installed electric baseboard heaters equipped with factory-installed receptacle outlets or outlets provided as a separate assembly by the manufacturer shall be permitted as the required outlet or outlets for the wall space utilized by such permanently installed heaters. Such receptacle outlets shall _____.

    A. be rated the same as the heater circuits
    B. be connected to and controlled by the heater circuits
    C. not be connected to the heater circuits
    D. be install above the heating elements

269. Service-entrance conductors to buildings or enclosures shall be installed shall not be smaller than _____ AWG unless in multiconductor cable. Multiconductor cable shall not be smaller than _____ AWG.

    A. 8, 10
    B. 4, 6
    C. 6, 8
    D. 6, 4

270. What is the maximum allowable ampacity for No. 6 aluminum conductors type THHN, 194 F rated insulation, installed in an ambient temperature of 30 C.

   A. 70
   B. 58.2
   C. 60.0
   D. 55

271. The grounding electrode conductor material selected shall be resistant to any _____ condition existing at the installation or shall be protected against corrosion.

   A. moisture
   B. high temperature
   C. corrosive
   D. below freezing

272. LFMC shall be securely fastened in place by an approved means within _____ in. of each box, cabinet, conduit body, or other conduit termination and shall be supported and secured at intervals not to exceed _____ ft. Exceptions ignored.

   A. 8, 4
   B. 12, 4 1/2
   C. 12, 6
   D. 18, 4.5

273. Each commercial building and each commercial occupancy accessible to pedestrians shall be provided with at least one outlet in an accessible location at each _____ to each tenant space for sign or outline lighting system use.

   A. exit
   B. store front
   C. window
   D. entrance

274. Where oil switches or air, oil, vacuum, or sulfur hexafluoride circuit breakers constitute a building disconnecting means, an isolating switch with _____ and meeting the requirements of 230.204(B), (C), and (D) shall be installed on the supply side of the disconnecting means and all associated equipment. Exception ignored.

    A. visible break contacts
    B. a explosion proof housing
    C. a hand operated lever
    D. non of the above

275. Type AC cable shall not be used _____.

    A. In damp or wet locations
    B. In air voids of masonry block or tile walls where such walls are exposed or subject to excessive moisture or dampness
    C. where subject to physical damage
    D. all the above
    E. none of the above

276. Rod and pipe grounding electrodes shall not be less than _____ ft in length and shall consist of the code specified materials.

    A. 4
    B. 6
    C. 8
    D. 10

277. For Fire alarm circuits, circuit integrity (CI) cables shall be supported at a distance not exceeding _____. Where located within _____ of the floor, as covered in 760.53(A)(1) and 760.130(1), as applicable, the cable shall be fastened in an approved manner at intervals of not more than _____. Cable supports and fasteners shall be steel

    A. 48 in. , 7 ft , 18 in.
    B. 24 in. , 2 ft , 18 in.
    C. 24 in. , 7 ft , 18 in.
    D. 24 in. , 7 ft , 36 in.

278. A warehouse storage area uses high pressure sodium lighting. What is the minimum lighting load in the warehouse area if it has 10,500 sq. ft. for storage.

    A. 2625
    B. 5262
    C. 3510
    D. 31500

279. Receptacles that provide power for water-pump motors or for other loads directly related to the circulation and sanitation system shall be located at least 10 ft from the inside walls of the pool, or not less than 6 ft from the inside walls of the pool if they meet which of the following conditions.

    A. Consist of single receptacles
    B. Employ a locking configuration
    C. Are of the grounding type
    D. Have GFCI protection
    E. all of the above

280. For luminaires near swimming pools, listed low-voltage luminaires not requiring grounding, not exceeding the low- voltage contact limit, and supplied by listed transformers or power supplies that comply with 680.23(A)(2) shall be per- mitted to be located less than _____ from the inside walls of the pool.

    A. 2 ft
    B. 5 ft
    C. 10 ft
    D. 15 ft

281. Up to three sets of _____ or two sets of 4-wire or 5-wire feeders shall be permitted to utilize a common neutral.

    A. 2-wire feeders
    B. insulated conductors
    C. 3-wire feeders
    D. 100 amp feeders

282. Conductors shall be permitted to be terminated based on the _____ °C temperature rating and ampacity as given in Table 310.60(C)(67) through Table 310.60(C)(86), unless otherwise identified.

   A. 60
   B. 75
   C. 90
   D. 100

283. What is the allowable ampacity for a No. 14 fixture wire.

   A. 15
   B. 17
   C. 20
   D. 25

284. Switches or circuit breakers shall not disconnect the _____ conductor of a circuit. Exception ignored.

   A. ungrounded
   B. energized
   C. grounded
   D. hot

285. Grounding electrode plates shall be installed not less than _____ in. below the surface of the earth.

   A. 45
   B. 95
   C. 30
   D. 60

286. Type MV cable shall be permitted for use on power systems rated up to and including _____ volts nominal.

   A. 600
   B. 2000
   C. 10,000
   D. 35,000

287. Any nonconductive paint, enamel, or similar coating shall be _____ at threads, contact points, and contact surfaces or be connected by means of fittings designed so as to make such removal unnecessary.

    A. removed
    B. tested
    C. continuity checked
    D. acid etched

288. Which of the following is not considered a special purpose branch circuit as outlined in the code.

    A. Circuits and equipment operating at less than 50 volts
    B. Cranes and hoists
    C. High pressure sodium parking lot lighting
    D. Fixed electric space-heating equipment

289. Exposed, normally non-current-carrying metal parts of fixed equipment supplied by or enclosing conductors or components that are likely to become energized shall be connected to an equipment grounding conductor under any of the following conditions:

    A. Where within 2.5 m (8 ft) vertically or 1.5 m (5 ft) horizontally of ground or grounded metal objects and subject to contact by persons
    B. Where located in a wet or damp location and not isolated
    C. Where in electrical contact with metal
    D. provided with GFI protectorWhere equipment operates with any terminal at over 150 volts to ground
    E. any of these

290. The lightning protection system ground terminals shall be bonded to the building or structure _____.

    A. lighting rod electrode
    B. frame
    C. foundation rebar
    D. grounding electrode system

291. On a 4-wire, delta-connected system where the midpoint of one phase winding is grounded, only the conductor or busbar having the higher phase voltage to ground shall be durably and permanently marked by an outer finish that is _____ in color or by other effective means.

    A. Red
    B. Yellow
    C. Orange
    D. Black or Red

292. For other than a totally enclosed switch- board or switchgear, a space not less than _____ shall be provided between the top of the switchboard or switchgear and any combustible ceiling, unless a noncom- bustible shield is provided between the switchboard or switchgear and the ceiling.

    A. 12 inches
    B. 1 ft.
    C. 3 ft
    D. none of the above

293. In a dwelling unit the back of a non-corner sink is located 14" from a wall, does the code require an outlet to be installed in that area if the distance to the nearest countertop outlet is 3 feet.

    A. Yes
    B. No

294. A reliable conductor to ensure the required electrical conductivity between metal parts required to be electrically connected.

    A. Branch Circuit
    B. Bonding Jumper
    C. Copper Wire
    D. Ground Wire

295. Where can additional grounding and bonding requirements for natural and artificially made bodies of water be found in the NEC.

   A. Article 517
   B. Article 230
   C. Article 690
   D. Article 682

296. Normally non-current-carrying conductive materials enclosing electrical conductors or equipment, or forming part of such equipment, shall be _____ so as to limit the voltage to ground on these materials.

   A. marked with a warning label
   B. insulated
   C. connected to earth
   D. ground fault protected

297. In hospitals the lighting demand factor of the _____ VA is 20 Percent.

   A. first 3000
   B. Remainder over 50,000
   C. VA Load from 3001 to 120,000
   D. all lighting

298. The required branch circuit copper conductor size for taps of a 15 amp rated circuit is No. _____.

   A. 18
   B. 14
   C. 12
   D. 10

299. At least _____ in. of free conductor, measured from the point in the box where it emerges from its raceway or cable sheath, shall be left at each outlet, junction, and switch point for splices or the connection of luminaries (fixtures) or devices. Exception ignored.

   A. 3
   B. 4
   C. 6
   D. 8

300. Service conductors passing over a roof shall be securely supported by substantial structures. Where practicable, such supports shall be _____.

   A. independent of the building
   B. coated with a non-conductive coating
   C. waterproof
   D. part of the buildings structure

301. In a 40 amp range circuit the ampere rating of a range receptacle shall be permitted to be _____ amps.

   A. 40
   B. 50
   C. 40 or 50
   D. all of the above

302. Conductors and cables in tunnels shall be located _____ and so placed or guarded to protect them from physical damage.

   A. above the tunnel floor
   B. under the tunnel floor
   C. within RNC
   D. in RMC

303. Overhead spans of open conductors and open multiconductor cables of not over 600 volts, nominal, shall have a clearance of not less than _____ ft for those areas listed in the 12-ft classification where the voltage exceeds 300 volts to ground.

   A. 10
   B. 12
   C. 15
   D. 18

304. What is the maximum number of conductors or fixture wires that can be used in electrical metallic tubing for No. 14 Type THHN conductors in trade size 27 tubing.

   A. 22
   B. 35
   C. 61
   D. 84

305. Each unit length of a heating cable a blue lead wire shall indicates what voltage requirement for that element.

   A. 120
   B. 208
   C. 240
   D. 277

306. Underground cable and conductors installed _____ shall be in a raceway. Exceptions ignored.

   A. in a basement
   B. under a building
   C. in an attic
   D. in a floor

307. A box or conduit body shall not be required where a luminaire (fixture) is used as a _____.

   A. junction box
   B. service panel
   C. grounding point
   D. raceway

308. In dwelling Units, all 125-volt, single-phase, 15- and 20-ampere receptacles installed in _____ shall have ground-fault circuit-interrupter protection for personnel. Exceptions ignored.

    A. Garages, and also accessory buildings that have a floor located at or below grade level not intended as habitable rooms and limited to storage areas, work areas, and areas of similar use.
    B. Bedrooms
    C. Bathrooms
    D. Both A and C

309. A conductor, other than a service conductor, that has overcurrent protection ahead of its point of supply that exceeds the value permitted for similar conductors that are protected as described elsewhere in 240.4.

    A. branch circuit conductor
    B. ungrounded conductor
    C. tap conductor
    D. overload conductor

310. Conductors normally used to carry current shall be of _____ unless otherwise provided in this Code and where the conductor material is not specified, the material and the sizes given in the Code shall apply to copper conductors.

    A. copper
    B. aluminum
    C. copper clad
    D. copper alloy

311. An assembly of a fuse support with either a fuse-holder, fuse carrier, or disconnecting blade. The fuseholder or fuse carrier may include a conducting element (fuse link) or may act as the disconnecting blade by the inclusion of a nonfusible member.

    A. Disconnect
    B. Cutout
    C. Breaker
    D. Fused Switch

312. Plate electrodes are permitted to be used for grounding electrodes if each plate electrode is expose not less than _____ sq. ft of surface to exterior soil.

    A. 2
    B. 4
    C. 6
    D. 9

313. The dwelling has a floor area of 1500 sq. ft. exclusive of an unfinished cellar not adaptable for future use, unfinished attic, and open porches. It has two 20-A small appliance circuits, one 20-A laundry circuit, two 4-kW wall-mounted ovens, one 5.1-kW counter-mounted cooking unit, a 4.5-kW water heater, a 1.2-kW dishwasher, a 5-kW combination clothes washer and dryer, six 7-A, 230-V room air-conditioning units, and a 1.5-kW permanently installed bathroom space heater. Assume wall-mounted ovens, counter-mounted cooking unit, water heater, dishwasher, and combination clothes washer and dryer kW ratings equivalent to kVA. What is the total air conditioner load in amperes.

    A. 36
    B. 42
    C. 50
    D. 18

314. In dwelling units, where two or more single-phase ranges are supplied by a 3-phase, 4-wire feeder or service, the total load shall be calculated on the basis of _____ the maximum number connected between any two phases.

    A. one and one half
    B. twice
    C. three times
    D. four times

315. What is the maximum number of conductors or fixture wires that can be used in electrical metallic tubing for No. 3 Type TW conductors in trade size 63 tubing.

    A. 8
    B. 10
    C. 15
    D. 20

316. In the installation and uses of electric wiring and equipment in ducts, plenums, and other air-handling spaces, _____ shall be installed in ducts or shafts containing only such ducts, used for vapor removal or for ventilation of commercial-type cooking equipment.

    A. no wiring systems of any type
    B. only rigid metallic conduit
    C. no flexible type conduit
    D. none of the above

317. A unit load of not less than _____ VA per square foot is required for general lighting in dwelling units.

    A. 1
    B. 2
    C. 3
    D. 4

318. A dwelling has a floor area of 1500 sq. ft, with unfinished cellar not adaptable for future use, unfinished attic, and open porches. Appliances are a 12-kW range and a 5.5-kW, 240-V dryer. Assume range and dryer kW ratings equivalent to kVA ratings in accordance with 220.54 and 220.55. What is the general lighting load in VA units.

    A. 2700 VA
    B. 3800 VA
    C. 4500 VA
    D. 6000 VA

319. A cable with insulated conductors enclosed within an overall, corrosion resistant, nonmetallic jacket describes what type of cable.

    A. Type SE
    B. Type NMC
    C. Type NMS
    D. Type UF

320. For feeders over 600 Volts, the ampacity of feeder conductors shall not be less than _____.

    A. 80 Amps
    B. 125% of the sum of the nameplate ratings of the transformers supplied when only transformers are supplied
    C. the sum of the nameplate ratings of the transformers supplied when only transformers are supplied
    D. 150% of the sum of the nameplate ratings of the transformers supplied when only transformers are supplied

321. Which of the following methods is not approved for the installation of service-entrance conductors.

    A. Electrical metallic tubing
    B. Busways
    C. Type NM cable
    D. Type MC cable

322. Two-wire dc circuits and ac circuits of two or more _____ shall be permitted to be tapped from the ungrounded conductors of circuits having a grounded neutral conductor. Switching devices in each tapped circuit shall have a pole in each ungrounded conductor.

    A. grounded conductors
    B. ungrounded conductors
    C. ungrounding conductors
    D. grounding conductors

323. A 30-ampere branch circuit shall be permitted to supply fixed lighting units with heavy-duty lampholders in other than a dwelling unit(s) or utilization equipment in any occupancy, and the rating of any one cord-and-plug-connected utilization equipment shall not exceed _____ percent of the branch-circuit ampere rating.
   A. 50
   B. 80
   C. 100
   D. 125

324. For conductors supplying a single motor, motor controller name-plate current ratings shall be permitted to be derived based on the _____ value of the motor current using an intermittent duty cycle and other control system loads, if present.

   A. phase
   B. intrinsic root squared
   C. rms
   D. meter measured

325. With respect to Solar Photovoltaic (PV) Systems equipment, where energy storage device input and output terminals are more than _____ from connected equipment, or where the circuits from these terminals pass through a wall or partition, the installation shall comply with the following: (1) A disconnecting means and overcurrent protection shall be provided at the energy storage device end of the circuit. Fused disconnecting means or circuit breakers shall be permitted to be used. (2) Where fused disconnecting means are used, the line terminals of the disconnecting means shall be con- nected toward the energy storage device terminals. (3) Overcurrent devices or disconnecting means shall not be installed in energy storage device enclosures where explosive atmospheres can exist. (4) A second disconnecting means located at the connected equipment shall be installed where the disconnecting means required by 690.71(H)(1) is not within sight of the connected equipment. (5) Where the energy storage device disconnecting means is not within sight of the PV system ac and dc discon- necting means, placards or directories shall be installed at the locations of all disconnecting means indicating the location of all disconnecting means.

   A. 3 ft
   B. 5 ft
   C. 10 ft
   D. 15 ft

326. For screw shell devices with attached leads, the conductor attached to the screw shell shall have a _____ finish and the outer finish of the other conductor shall be of a solid color that will not be confused the grounded conductor terminal.

   A. green
   B. black
   C. gold
   D. white or gray

327. Overhead spans of open conductors and open multiconductor cables of not over 1000 volts, nominal, shall have a clearance of not less than _____ ft above finished grade, sidewalks, or from any platform or projection from which they might be reached where the voltage does not exceed 150 volts to ground and accessible to pedestrians only

   A. 10
   B. 12
   C. 15
   D. 18

328. Where a branch circuit supplies continuous loads or any combination of continuous and noncontinuous loads, the minimum branch-circuit conductor size, shall have an allowable ampacity not less than the noncontinuous load plus _____ percent of the continuous load. Exceptions ignored.

   A. 50
   B. 75
   C. 100
   D. 125

329. An assembly of two or more single-pole fuses.

   A. Main Disconnect
   B. Multiple Fuse
   C. Gang Disconnect
   D. Parallel Breaker

330. What is the demand factor percentage for 9 household electric clothes dryers in a multi-family dwelling unit.

   A. 100
   B. 85
   C. 65
   D. 55

331. A device that establishes a connection between two or more conductors or between one or more conductors and a terminal by means of mechanical pressure and without the use of solder.

   A. Solder Connector
   B. Type P Connector
   C. Pressure Connector
   D. Wire Splice

332. For a one-family dwelling, the feeder disconnecting means shall have a rating of not less than _____ amperes, 3-wire.

   A. 60
   B. 90
   C. 100
   D. 125

333. Underground wiring shall not be permitted under a pool or within the area extending _____ ft horizontally from the inside wall of the pool unless this wiring is necessary to supply pool equipment.

   A. 3
   B. 5
   C. 8
   D. 10

334. A manually operated device used in conjunction with a transfer switch to provide a means of directly connecting load conductors to a power source and of disconnecting the transfer switch.

    A. Bypass Isolation Switch
    B. Transfer Switch
    C. Double Pole Switch
    D. 4 Way Switch

335. An energy management system shall not override the load shedding controls put in place to ensure the minimum electrical capacity. Which of the following systems are not covered under this requirement.

    A. Fire pumps
    B. Emergency systems
    C. Legally required standby systems
    D. Critical operations power systems
    E. none of them

336. One method of marking an insulated grounded conductor larger than 6 AWG shall be _____.

    A. by using a bare copper conductor
    B. by noting which terminal to which it is connected
    C. At the time of installation, by a distinctive white or gray marking at its terminations. This marking shall encircle the conductor or insulation.
    D. by a distinctive green marking encircling the conductor or insulation at its terminations installed at the time of installation.

337. The normally non-current-carrying metal parts of all service enclosures shall be effectively _____.

    A. bonded together
    B. insulated
    C. protected
    D. locked

338. Cellular Concrete Floor Raceways shall not be used In commercial garages, other than for supplying _____ outlets or extensions to the area below the floor but not above.

   A. floor
   B. wall
   C. ceiling
   D. roof

339. A dwelling has a floor area of 2000 sq. ft exclusive of an unfinished cellar not adaptable for future use, unfinished attic, and open porches. It has a 12-kW range, a 4.5-kW water heater, a 1.2-kW dishwasher, a 5-kW clothes dryer, and a 2 1/2-ton (24-A) heat pump with 15 kW of backup heat. What is the total heating and cooling load in VA units.

   A. 5,760
   B. 15,000
   C. 20,760
   D. 15,510

340. A 6 AWG grounding electrode conductor that is free from exposure to physical damage shall be permitted to be run along the surface of the building construction without metal covering or protection where it is _____.

   A. a stranded bare copper conductor
   B. a solid, insulated copper conductor
   C. a stranded, insulated copper conductor
   D. securely fastened to the construction

341. The length of the cord for a kitchen waste disposer shall not be less than _____ in. and not more than _____ in.

   A. 24, 42
   B. 30, 48
   C. 36, 72
   D. 18, 36

342. Lighting accessory cords on the load side of a listed _____ shall not be required to contain an equipment grounding conductor.

    A. Class 2 power source
    B. Class 1 power source
    C. Isolation transformer
    D. All of the above

343. All switches and circuit breakers used as switches shall be located so that they may be operated from a readily accessible place and shall be installed such that the center of the grip of the operating handle of the switch or circuit breaker, when in its highest position, is not more than _____ above the floor or working platform. Exceptions ignored.

    A. 5 ft 2 in.
    B. 4 ft 7 in.
    C. 6 ft 6 in.
    D. 6 ft 7 in.

344. Where connected to a branch circuit having a rating in excess of 20 amperes, lampholders shall be of _____ and such lampholder shall have a rating of not less than 660 watts if of the admedium type, or not less than 750 watts if of any other type.

    A. high voltage
    B. high amperage
    C. heavy-duty type
    D. rough use

345. Direct-buried conductors or cables shall be permitted to be spliced or tapped _____.

    A. without the use of splice boxes
    B. only with the use of splice boxes
    C. only with the use of waterproof splice boxes
    D. and tapped with a minimum of 1/2" of friction tape

346. The use of EMT shall be permitted for _____ work.

    A. underwater
    B. exposed
    C. concealed
    D. both B and C

347. Which of the following areas of construction does the NEC (National Electrical Code) not cover?

    A. private premises
    B. parking lots
    "C. automotive vehicles other than mobile homes and recreational vehicles"
    D. Installations used by the electric utility

348. The equipment grounding conductor run with or enclosing the circuit conductors may in the form of a wire or a busbar _____.

    A. of any shape
    B. within the cable or conduit
    C. of any metallic element
    D. of circular copper or aluminum

349. Ground-fault circuit-interrupter protection for personnel shall be provided for outlets that supply boat hoists installed in dwelling unit locations and supplied by _____.

    A. all 125-volt, 15- and 20-ampere or greater branch circuits
    B. all 20-ampere branch circuits
    C. outlets not exceeding 240 volts
    D. three wire branch circuits 125-volt, 15- and 20-ampere

350. Wind turbines shall be required to have a readily accessible manual shutdown button or switch. Operation of the button or switch shall result in a parked turbine state that shall either stop the turbine rotor or allow limited rotor speed combined with a means to de-energize the turbine output circuit. Exception: Turbines with a swept area of less than _____ shall not be required to have a manual shutdown button or switch.

    A. 100 sq ft
    B. 250 sq ft
    C. 538 sq ft
    D. 1000 sq ft

351. Multiconductor cables used for overhead service conductors shall be attached to buildings or other structures by fittings identified for use with service conductors. Open conductors shall be attached to fittings identified for use with service conductors or to noncombustible, _____ insulators securely attached to the building or other structure.

    A. glass
    B. metallic
    C. plastic
    D. nonabsorbent

352. Where the building or structure disconnecting means does not disconnect the _____ conductor from the _____ conductors in the building or structure wiring, other means shall be provided for this purpose at the location of disconnecting means. A terminal or bus to which all grounded conductors can be attached by means of pressure connectors shall be permitted for this purpose.

    A. grounded
    B. ungrounded
    C. grounding
    D. uninsulated

353. Receptacles and cord connectors shall be rated not less than _____ amperes, 125 volts, or 250 volts, and shall be of a type not suitable for use as lampholders.

   A. 12
   B. 15
   C. 20
   D. 25

354. IMC shall be securely fastened within _____ ft of each outlet box, junction box, device box, cabinet, conduit body, or other conduit termination. Fastening shall be permitted to be increased to a distance of _____ ft where structural members do not readily permit fastening at the standard minimum distance. Exceptions ignored.

   A. 5, 10
   B. 4, 10
   C. 4, 8
   D. 3, 5

355. What is the maximum fill allowance for a 4 inch square box 2.125" deep, for No. 12 copper conductors.

   A. 10
   B. 12
   C. 13
   D. 15

356. Where a service mast is used for the support of service-drop conductors, it shall be of adequate strength or be supported by _____ to withstand safely the strain imposed by the service drop.

   A. lateral cables
   B. angle ties
   C. rigid metallic tubing
   D. braces or guys

100

357. All non-current-carrying metal parts of electric equipment and all metal raceways and cable sheaths shall be solidly grounded and bonded to all metal pipes and rails at the portal and at intervals not exceeding _____ ft throughout a tunnel.

    A. 500
    B. 1000
    C. 1500
    D. 2000

358. In both exposed and concealed locations, where a cable or raceway type wiring method is installed through bored holes in joists, rafters, or wood members, holes shall be bored so that the edge of the hole is not less than _____ in. from the nearest edge of the wood member. Where this distance cannot be maintained, the cable or raceway shall be protected from penetration by screws or nails by a steel plate or bushing, at least _____ in. thick, and of appropriate length and width installed to cover the area of the wiring. Exceptions ignored.

    A. 1.5, .125
    B. 1.625, .20
    C. 1.25, .0625
    D. 1.75, .25

359. Where a building or other structure that is served by a branch circuit or feeder on the load side of a service disconnecting means shall be supplied by only _____. Ignoring all special conditions.

    A. only one feeder or branch circuit
    B. a separate service
    C. multiple branch circuit feeders
    D. new service entrance cabling

360. A factory assembly of one or more insulated conductors with an integral or an overall covering of nonmetallic material suitable for direct burial in the earth.

    A. Type NM
    B. Type SE
    C. Type UN
    D. Type UF

361. Where used outside, aluminum or copper-clad aluminum grounding electrode conductors shall not be terminated within _____ in. of the earth.

    A. 12
    B. 18
    C. 24
    D. 36

362. Surge arresters installed in accordance with the requirements of Article 280 shall be permitted on each _____ overhead service conductor.

    A. ungrounded
    B. grounded
    C. grounding
    D. bonding

363. A unit load of not less than _____ VA per square foot is required for general lighting in Restaurant occupancies.

    A. 1
    B. 2
    C. 3
    D. 4

364. Taps from bare conductors shall leave the gutter _____ their terminal connections, and conductors shall not be brought in contact with uninsulated current-carrying parts of different potential.

    A. opposite
    B. parallel to
    C. within 3 inches of
    D. bare from

365. What is the minimum bending radius in inches for 2" EMT using a One Shot bender.

    A. 7
    B. 8.25
    C. 9.5
    D. 11

366. What is the maximum allowable ampacity for No. 2/0 copper conductors type THHN, 167 F rated insulation, installed in an ambient temperature of 30 C. There are 2 current carrying conductors in the cable.

    A. 135
    B. 150
    C. 115
    D. 195

367. Raceways on exteriors of buildings or other structures shall be arranged to drain and shall be _____ in wet locations.

    A. waterproof
    B. raintight
    C. rainproof
    D. suitable for use

368. Overhead spans of open conductors and open multiconductor cables of not over 600 volts, nominal, shall have a clearance of not less than _____ ft over residential property and driveways, and those commercial areas not subject to truck traffic where the voltage does not exceed 300 volts to ground.

    A. 10
    B. 12
    C. 14
    D. 18

369. In a multiwire branch circuit all conductors shall originate from the same _____ or similar distribution equipment.

    A. location
    B. panelboard
    C. junction box
    D. meter

370. Aluminum, copper-clad aluminum, or copper conductors of size 1/0 AWG and larger, comprising each phase, polarity, neutral, or grounded circuit conductor, shall be permitted to be connected in _____ (electrically joined at both ends). Exceptions Ignored.

    A. parallel
    B. series
    C. delta
    D. wye

371. In locations where electric equipment is likely to be exposed to physical damage, enclosures or _____ shall be so arranged and of such strength as to prevent such damage.

    A. insulation
    B. guards
    C. conductors
    D. equipment

372. Bends shall be made so that the tubing is not damaged and the internal diameter of the tubing is not effectively _____.

    A. increased
    B. reduced
    C. lengthened
    D. non of the above

373. To prevent corrosion, raceways, cable trays, cablebus, auxiliary gutters, cable armor, boxes, cable sheathing, cabinets, elbows, couplings, fit-tings, supports, and support hardware shall be of materials suitable for the _____ in which they are to be installed.

    A. location
    B. position
    C. environment
    D. humidity

374. As pretaining to class 1, class 2, and class 3 remote-control, signaling, and power-limited circuits, separation of ground-fault protection time-current characteristics shall conform to the manufacturer's recommendations and shall consider all required tolerances and disconnect operating time to achieve _____ selectivity.

   A. 50 percent
   B. 75 percent
   C. 100 percent
   D. none of these

375. The disconnecting means for each supply shall consist of not more than _____ switches or _____ circuit breakers mounted in a single enclosure, in a group of separate enclosures, or in or on a switchboard or switchgear.

   A. six, six
   B. two, two
   C. three, three
   D. eight, eight

376. At least one 125-volt, single- phase, 15- or 20-ampere-rated receptacle outlet shall be installed within 450 mm (18 in) of the top of a show window for each _____ or major fraction thereof of show window area measured horizontally at its maximum width.

   A. 6 linear ft
   B. 12 linear ft
   C. 10 linear ft
   D. 8 linear ft

377. Two or more grounding electrodes that are effectively bonded together shall be considered _____.

   A. a multiple grounding electrode system
   B. a continuous grounding electrode system.
   C. a single grounding electrode system
   D. a redundant grounding electrode system.

378. Cabinets and cutout boxes shall have approved space to accommodate all conductors installed in them without _____.

    A. bending
    B. crowding
    C. kinking
    D. trimming

379. The point of attachment of the overhead service conductors to a building or other structure shall provide the minimum clearances an in no case shall this point of attachment be less than _____ ft above finished grade.

    A. 8
    B. 10
    C. 12
    D. 14

380. Conduits or raceways through which moisture may contact live parts shall be _____.

    A. RMC
    B. RMC or RNC
    C. EMT, RMC, or RNC
    D. sealed or plugged at either or both ends

381. Bare _____ grounding electrode conductors shall not be used where in direct contact with masonry or the earth or where subject to corrosive conditions.

    A. aluminum
    B. copper
    C. copper-clad aluminum
    D. aluminum or copper-clad aluminum

382. An insulated conductor that is intended for use as a grounded conductor, where contained within a flexible cord, shall be identified by a _____ outer finish or by methods permitted by 400.22.

    A. white or gray
    B. green
    C. bare
    D. varnished

383. The maximum current, in amperes, that a conductor can carry continuously under the conditions of use without exceeding its temperature rating.

    A. average load
    B. standard operating current
    C. maxium power
    D. ampacity

384. The feeder conductor ampacity shall not be less than that of the _____ where the feeder conductors carry the total load supplied by service conductors with an ampacity of 55 am- peres or less.

    A. the total load
    B. 4/0
    C. No. 8
    D. service conductors

385. Locations of lamps for outdoor lighting shall be below all energized conductors, transformers, or other electric utilization equipment, unless either of the following apply

    A. either C or D
    B. Equipment is controlled by a disconnecting means that can be locked in the closed position.
    C. Equipment is controlled by a disconnecting means that is lockable in accordance with 110.25.
    D. Clearances or other safeguards are provided for relamping operations.

386. Conductors, other than _____, shall be protected against overcurrent in accordance with their ampacities.

    A. service lateral conductors
    B. heating overload conductors
    C. flexible cords, flexible cables, and fixture wires
    D. variable gauge conductors

387. A compartment or chamber to which one or more air ducts are connected and that forms part of the air distribution system.

   A. Heat Exchange
   B. Plenum
   C. Duct
   D. Air Vent

388. Sheet metal auxiliary gutters shall be supported and secured throughout their entire length at intervals not exceeding _____ ft.

   A. 3
   B. 5
   C. 6
   D. 8

389. In dwelling units, at least one receptacle outlet shall be installed in bathrooms within _____ of the outside edge of each basin and the receptacle outlet shall be located on a wall or partition that is adjacent to the basin or basin countertop, located on the countertop, or installed on the side or face of the basin cabinet not more than 300 mm (12 in.) below the countertop. Receptacle outlet assemblies listed for the application shall be permitted to be installed in the countertop

   A. 3 ft
   B. 6 ft
   C. 4 ft
   D. 18 in.

390. Throughout the Code, the voltage considered shall be that at which the circuit operates, and the voltage rating of electrical equipment shall not be less than the _____ of a circuit to which it is connected.

   A. under current protection rating
   B. over current protection rating
   C. nominal voltage
   D. stable effective voltage

391. Equipment intended to interrupt current at fault levels shall have an interrupting rating at nominal circuit _____ sufficient for the current that is available at the line terminals of the equipment.

   A. voltage
   B. power
   C. resistance
   D. size

392. "Completed wiring installations shall be free from short circuits and from _____ other than as required or permitted in Article 250.

Completed wiring installations shall be free from short circuits, _____, or any connections to ground other than as required or permitted elsewhere in this Code."

   A. switches
   B. ground faults
   C. fuses
   D. teminals

393. In a one family dwelling and each unit of a two-family dwelling that is at grade level, at least one receptacle outlet readily accessible from grade and not more than _____ above grade shall be installed at the front and back of the dwelling.

   A. 6 1/2 ft
   B. 78 in.
   C. 6.5 feet
   D. all of the above

394. Equipment that has an open-circuit voltage exceeding _____ volts shall not be installed in or on dwelling occupancies.

   A. 208
   B. 240
   C. 1000
   D. 1200

395. Where the load is calculated on the basis of volt-amperes per square meter or per square foot, the wiring system up to and including the branch-circuit panelboard(s) shall be provided to serve not less than the calculated load. This load shall be _____ among multioutlet branch circuits within the panelboard(s). Branch-circuit overcurrent devices and circuits shall only be required to be installed to serve the connected load.

    A. connected in parallel
    B. concentrated
    C. evenly proportioned
    D. determined and calculated

396. Overcurrent devices shall be readily accessible and shall be installed so that the center of the grip of the operating handle of the switch or circuit breaker, when in its highest position, is not more than 6 ft ___ in. above the floor or working platform. Exceptions ignored.

    A. 6
    B. 7
    C. 8
    D. 10

397. Type NM cables shall be durably marked on the surface, the AWG size or circular mil area shall be repeated at intervals not exceeding _____ in., and all other markings shall be repeated at intervals not exceeding _____ in..

    A. 12, 24
    B. 18, 48
    C. 24, 40
    D. 40, 24

398. Where a feeder supplies branch circuits in which equipment grounding conductors are required, the feeder shall include or provide a(n) _____, in accordance with the provisions of 250.134, to which the equipment grounding conductors of the branch circuits shall be connected.

    A. water pipe
    B. chase connection
    C. grounded earth electrode
    D. equipment grounding conductor

399. RMC threadless couplings and connectors used with conduit shall be made tight. Where buried in masonry or concrete, they shall be the type.

   A. watertight
   B. airtight
   C. waterproof
   D. concrete tight

400. Where the assembly, including the over-current devices protecting the feeder(s), is listed for operation at _____ percent of its rating, the ampere rating of the overcurrent device shall be permitted to be not less than the sum of the continuous load plus the noncontinuous load.

   A. 80
   B. 125
   C. 100
   D. 150

401. A 15- or 20-ampere branch circuit shall be permitted to supply lighting units or other utilization equipment, or a combination of both. Exceptions ignored.

   A. This is True
   B. This is False

402. Operation of equipment in excess of normal, full-load rating, or of a conductor in excess of rated ampacity that, when it persists for a sufficient length of time, would cause damage or dangerous overheating. A fault, such as a short circuit or ground fault, is not an overload.

   A. Ground Fault
   B. Arc Fault
   C. Overload
   D. Short Circuit

403. A device or group of devices that serves to govern, in some predetermined manner, the electric power delivered to the apparatus to which it is connected.

    A. Remote Relay
    B. Controller
    C. Circuit Breaker
    D. Governor

404. The general calculated load shall be not less than _____ percent of the first 10 kVA plus _____ percent of the remainder of the loads.

    A. 100, 125
    B. 80, 100
    C. 100, 40
    D. 90, 50

405. A receptacle outlet shall be installed wherever flexible cords with attachment plugs are used, and where flexible cords are permitted to be permanently connected, receptacles shall be _____ for such cords.

    A. permitted to be omitted
    B. installed
    C. permanently installed
    D. marked

406. The NEC (National Electrical Code) provided that the number of wires and circuits confined in a single enclosure be _____. Limiting the number of circuits in a single enclosure minimizes the effects from a short circuit or ground fault in one circuit.

    A. varyingly restricted
    B. restricted to 3 circuits
    C. restricted to 1 circuit
    D. unrestricted

407. A grounding terminal or grounding-type device can _____.

    A. be used by grounded conductor
    B. bond the grounded conductor to the neutral
    C. not be used for purposes other than grounding
    D. be used in lieu of a GFI

408. In dwelling unit garages. In each attached garage and in each detached _____. The branch circuit supplying this receptacle(s) shall not supply outlets outside of the garage. At least one receptacle outlet shall be installed for each car space.

    A. detached garage or accessory building with electric power
    B. walkway to the garage
    C. 6' space along the floorline
    D. 12' or wider carport

409. In feeders over 600 Volts Supplying Transformers and Utilization Equipment, the ampacity of feeders supplying a combination of transformers and utilization equipment shall not be less than the sum of the nameplate ratings of the transformers and 125 percent of the designed potential load of the utilization equipment that will be operated _____.

    A. independently
    B. non continuously
    C. simultaneously
    D. continuously

410. Switches and circuit breakers used as disconnecting means shall be of the _____ type.

    A. toggle
    B. T
    C. indicating
    D. grounded disconnect conductor

411. The nominal voltage of branch circuits shall not exceed _____ nominal, between conductors In dwelling units and guest rooms or guest suites of hotels, motels, and similar occupancies that supply the terminals of luminaries (lighting fixtures) and cord-and-plug-connected loads 1440 volt-amperes, nominal, or less or less than 1/4hp.

    A. 400
    B. 220
    C. 120
    D. 110

412. Conductors other than _____ shall not be installed in the same service raceway or service cable. Exceptions ignored.

    A. copper
    B. service conductors
    C. ungrounded
    D. SE or USE

413. Rod and pipe grounding electrodes shall be installed such that at least _____ ft of length is in contact with the soil.

    A. 12
    B. 10
    C. 8
    D. 6

414. The minimum branch-circuit conductor size shall have an allowable ampacity not less than the _____ load to be served after the application of any adjustment or correction factors

    A. maximum
    B. minimum
    C. constant
    D. intermittent

415. For cord-and-plug-connected equipment in other than storable pools, the flexible cords shall not exceed _____ ft in length.

   A. 3
   B. 6
   C. 10
   D. 15

416. Branch circuits larger than 50 amperes shall supply only _____ outlet loads.

   A. non-lighting
   B. lighting
   C. ranges
   D. cooking equipment

417. The calculated lighting load for an office building is 24,500 VA. What is the maximum sq ft that the office could have if this were the lighting load.

   A. 4500
   B. 5600
   C. 7000
   D. 8300

418. In walls of concrete, tile, or other noncombustible material, cabinets shall be installed so that the front edge of the cabinet is not set back of the finished surface more than _____ in.

   A. 1/8
   B. 1/4
   C. 3/8
   D. 1/2

419. _____ used in the construction of a luminaire (fixture) box.

   A. Screws, crimps or solder may be
   B. Only silver solder may be
   C. No solder shall be
   D. none of the above

420. There shall not be more than the equivalent of four quarter bends _____ degrees total between pull points, for example, conduit bodies and boxes.

    A. 180
    B. 270
    C. 360
    D. 450

421. An insulated grounded conductor of 6 AWG or smaller shall be identified by _____.

    A. A continuous white outer finish.
    B. A continuous gray outer finish.
    C. Three continuous white or gray stripes along the conductor's entire length on other than green insulation.
    D. Wires that have their outer covering finished to show a white or gray color but have colored tracer threads in the braid identifying the source of manufacture shall be considered as meeting the provisions of this section.
    E. Any of the above and more options are available

422. For attics and underfloor spaces containing equipment requiring servicing, such as heating, air-conditioning, and refrigeration equipment, at least one _____ containing a switch or controlled by a wall switch shall be installed in such spaces. At least one point of control shall be at the usual point of entry to these spaces.

    A. lighting outlet
    B. power control
    C. junction box
    D. panelboard

423. What is the minimum distance between bare metal parts, busbars, and so forth in a panelboard between live parts to ground where the voltage 240 V.

    A. 1 inch
    B. 3/4 inch
    C. 1/2 inch
    D. 1/4 inch

424. Maximum Number of Disconnects, Two or three single-pole switches or breakers capable of individual operation shall be permitted on multiwire circuits, one pole for each ungrounded conductor, as one multipole disconnect, provided they are equipped with identified handle ties or a master handle to disconnect all ungrounded conductors with no more than _____ operations of the hand.

    A. six
    B. twelve
    C. four
    D. eight

425. A unit load of not less than _____ VA per square foot is required for general lighting in Hospital occupancies.

    A. 1
    B. 2
    C. 3
    D. 4

426. Wire connectors or splicing means installed on conductors for direct burial shall _____.

    A. be listed for such use
    B. not be used
    C. encased in concrete
    D. made waterproof

427. Noncombustible surfaces that are broken or incomplete shall be repaired so there will be no gaps or open spaces greater than _____ in. at the edge of the cabinet or cutout box employing a flush-type cover.

    A. 1/16
    B. 1/8
    C. 3/16
    D. 1/4

428. What is the ampacity rating of No. 4 (AWG) copper conductors at 75 C. in multiconductor UF cables with not more than three insulated conductors, rated 0 through 2000 Volts, in free air based on ambient air temperature of 40°C for type UF, and USE cables

    A. 69
    B. 89
    C. 100
    D. 104

429. Receptacle outlets in or on floors shall not be counted as part of the required number of receptacle outlets unless located within _____ of the wall.

    A. 1 ft.
    B. 1.5 ft.
    C. 2 ft.
    D. 8 in.

430. Nonmetallic boxes shall be permitted only with open wiring on insulators, _____, cabled wiring methods with entirely nonmetallic sheaths, flexible cords, and nonmetallic raceways. Exceptions ignored.

    A. electrical metallic tubing
    B. concealed knob-and-tube wiring
    C. rigid metallic tubing
    D. IMT

431. Circuits not exceeding 120 volts, nominal, between conductors shall be permitted to supply the following which of the following:

    A. The terminals of lampholders applied within their voltage ratings
    B. Cord-and-plug-connected or permanently connected utilization equipment
    C. The terminals of lampholders applied within their voltage ratings
    D. Auxiliary equipment of electric-discharge lamps
    E. All of the above

432. A device designed to open and close a circuit by nonautomatic means and to open the circuit automatically on a predetermined overcurrent without damage to itself when properly applied within its rating.

    A. Fuse
    B. Circuit Breaker
    C. Cartridge Fuse
    D. Fusible Link

433. Where a submersible pump is used in a metal well casing, the well casing shall be _____ to the pump circuit equipment grounding conductor.

    A. tied
    B. clamped
    C. bonded
    D. connected

434. In dwelling units, at least one _____ outlet shall be installed in every habitable room and bathroom. Exceptions ignored.

    A. appliance
    B. wall switch-controlled lighting
    C. hair dryer
    D. 20 amp.

435. What is the maximum percent of EMT conduit total interior area that can be filled if more that 2 conductors are used in the tube.

    A. 53
    B. 31
    C. 40
    D. 60

Section 2

Timed Exams

Each exam has 80 Questions, there are 5 exams with the questions in random order.
Time yourself, allow yourself 3 hours to complete each exam. This will develop your time management skills as well as giving you review knowledge and lookup practice.

Timed Exam 1   31/2 Hours to Complete

Timed Exam 1-1   What is the minimum bending radius in inches for 2" EMT using a One Shot bender.

      A. 7
      B. 8.25
      C. 9.5
      D. 11

Timed Exam 1-2   Wind turbines shall be required to have a readily accessible manual shutdown button or switch. Operation of the button or switch shall result in a parked turbine state that shall either stop the turbine rotor or allow limited rotor speed combined with a means to de-energize the turbine output circuit. Exception: Turbines with a swept area of less than _____ shall not be required to have a manual shutdown button or switch.

      A. 100 sq ft
      B. 250 sq ft
      C. 538 sq ft
      D. 1000 sq ft

Timed Exam 1-3   Overhead spans of open conductors and open multiconductor cables of not over 600 volts, nominal, shall have a clearance of not less than _____ ft for those areas listed in the 12-ft classification where the voltage exceeds 300 volts to ground.

      A. 10
      B. 12
      C. 15
      D. 18

Timed Exam 1-4   No parts of cord-connected luminaries (fixtures), chain-, cable-, or cord-suspended-luminaries (fixtures), lighting track, pendants, or ceiling-suspended (paddle) fans shall be located within a zone measured _____ ft horizontally and _____ ft vertically from the top of the bathtub rim or shower stall threshold.

        A. 3, 8
        B. 5, 8
        C. 5, 10
        D. 8, 8

Timed Exam 1-5   Continuous Load is a load where the maximum current is expected to continue for ____ hours or more.

        A. 1
        B. 2
        C. 3
        D. 4

Timed Exam 1-6   What is the maximum number of conductors or fixture wires that can be used in electrical metallic tubing for No. 14 Type THHN conductors in trade size 27 tubing.

        A. 22
        B. 35
        C. 61
        D. 84

Timed Exam 1-7   The grounding electrode shall be _____ the grounding electrode conductor connection to the system.

        A. within 10 feet of
        B. within the same general area but no case more than 30' from
        C. as near as practicable to, and preferably in the same area as,
        D. directly adjacent to

Timed Exam 1-8   Admitting close approach; not guarded by locked doors, elevation, or other effective means.

       A. approve
       B. accessible
       C. workable
       D. open

Timed Exam 1-9   A 3000 sq. ft. store, has 30 ft of show window. There are a total of 80 duplex receptacles. The service is 120/240 V, single phase 3-wire service. Actual connected lighting load is 8500 VA. What is the total calculated receptacle load in VA

       A. 14,400
       B. 10,000
       C. 2,200
       D. 12,200

Timed Exam 1-10   In a Cellular Metal Floor Raceway the combined cross-sectional area of all conductors or cables shall not exceed _____ percent of the interior cross-sectional area of the cell or header.

       A. 25
       B. 40
       C. 60
       D. 75

Timed Exam 1-11   What is the minimum distance between bare metal parts, busbars, and so forth in a panelboard between live parts to ground where the voltage 240 V.

       A. 1 inch
       B. 3/4 inch
       C. 1/2 inch
       D. 1/4 inch

Timed Exam 1-12   An assembly of two or more single-pole fuses.

    A. Main Disconnect
    B. Multiple Fuse
    C. Gang Disconnect
    D. Parallel Breaker

Timed Exam 1-13   A cable and components used to ensure survivability of critical circuits for a specified time under fire conditions

    A. Asbestos Shielded Cable System
    B. Gemini Duplicated Cable System
    C. Refractory Shielded Cable Sytem
    D. Fire-Resistive Cable System.

Timed Exam 1-14   Laundry Circuit Load. A load of not less than 1500 volt-amperes shall be included for each 2-wire laundry branch circuit installed, this load shall be permitted to be included with the _____ load and subjected to the demand factors provided in Table 220.42.

    A. overall appliance
    B. utility circuit
    C. heating
    D. general lighting

Timed Exam 1-15   The required branch circuit copper conductor size for taps of a 30 amp rated circuit is No. _____.

    A. 14
    B. 12
    C. 10
    D. 8

Timed Exam 1-16   Where space limitations prevent wiring from being routed a distance 5 ft or more from a pool, such wiring shall be permitted where installed in complete raceway systems of _____.

    A. rigid metal conduit
    B. intermediate metal conduit
    C. or a nonmetallic raceway system
    D. none of the above
    E. all of the above

Timed Exam 1-17   A device, group of devices, or other means whereby the conductors of a circuit can be disconnected from their source of supply.

       A. Switch
       B. Disconnecting Means
       C. Closed Circuit Switch
       D. Relay Switch

Timed Exam 1-18   Expansion fittings and telescoping sections of metal raceways shall be made electrically continuous by _____ or other means.

       A. application of oxidation inhibitors
       B. cleaning bright before assembly
       C. equipment bonding jumpers
       D. the use of bare grounding conductors

Timed Exam 1-19   A snap switch shall not be grouped or ganged in enclosures with other snap switches, receptacles, or similar devices, unless they are arranged so that the voltage between adjacent devices does not exceed _____ volts, or unless they are installed in enclosures equipped with identified, securely installed barriers between adjacent devices.

       A. 120
       B. 240
       C. 300
       D. 600

Timed Exam 1-20   Wiring located within the cavity of a fire-rated floor-ceiling or roof-ceiling assembly shall be secured to, or supported by, an independent means of secure support.   Where independent support wires are used, they shall be _____.   Exceptions ignored.

       A. galvanized steel No. 10 or larger.
       B. steel, copper, brass or aluminum
       C. distinguishable by twists within the bottom 12 inches of the wires
       D. distinguishable by color, tagging, or other effective means from those that are part of the fire-rated design

Timed Exam 1-21 Nonmetallic boxes shall be permitted only with open wiring on insulators, _____, cabled wiring methods with entirely nonmetallic sheaths, flexible cords, and nonmetallic raceways. Exceptions ignored.

    A. electrical metallic tubing
    B. concealed knob-and-tube wiring
    C. rigid metallic tubing
    D. IMT

Timed Exam 1-22 Raceways, cable assemblies, boxes, cabinets, and fittings shall be securely fastened in place. Support wires that do not provide secure support shall not be _____.

    A. permitted
    B. allowed to be counted as part of the structural system of fastening
    C. permitted as the sole support
    D. copper or aluminum

Timed Exam 1-23 Circuits that supply electric motor driven fire pumps shall be supervised from _____ disconnection.

    A. inadvertent
    B. overload
    C. water load
    D. all of the above

Timed Exam 1-24   Ground-Fault Circuit Interrupter (GFCI)

      A. A system or circuit conductor that is intentionally grounded.
      B. An unintentional, electrically conducting connection between an ungrounded conductor of an electrical circuit and the normally non-current-carrying conductors, metallic enclosures, metallic raceways, metallic equipment, or earth.
      C. All circuit conductors between the service equipment, the source of a separately derived system, or other power supply source and the final branch-circuit overcurrent device.
      D. A device intended for the protection of personnel that functions to deenergize a circuit or portion thereof within an established period of time when a current to ground exceeds the values established for a Class A device.

Timed Exam 1-25   The dwelling has a floor area of 1500 sq. ft. exclusive of an unfinished cellar not adaptable for future use, unfinished attic, and open porches. It has two 20-A small appliance circuits, one 20-A laundry circuit, two 4-kW wall-mounted ovens, one 5.1-kW counter-mounted cooking unit, a 4.5-kW water heater, a 1.2-kW dishwasher, a 5-kW combination clothes washer and dryer, six 7-A, 230-V room air-conditioning units, and a 1.5-kW permanently installed bathroom space heater. Assume wall-mounted ovens, counter-mounted cooking unit, water heater, dishwasher, and combination clothes washer and dryer kW ratings equivalent to kVA. What is the total air conditioner load in amperes.

      A. 36
      B. 42
      C. 50
      D. 18

Timed Exam 1-26   Installations underground or in concrete slabs or masonry in direct contact with the earth; in locations subject to saturation with water or other liquids, such as vehicle washing areas; and in unprotected locations exposed to weather.

      A. Wet Location
      B. Dry Location
      C. Damp Location
      D. High Humidity Location

Timed Exam 1-27   The width of the working space in front of the electric equipment shall be the width of the equipment or _____ in., whichever is greater and the work space shall permit at least a _____ degree opening of equipment doors or hinged panels.

        A. 90, 30
        B. 45, 60
        C. 30, 90
        D. 60, 45

Timed Exam 1-28   Noncombustible surfaces that are broken or incomplete shall be repaired so there will be no gaps or open spaces greater than _____ in. at the edge of the cabinet or cutout box employing a flush-type cover.

        A. 1/16
        B. 1/8
        C. 3/16
        D. 1/4

Timed Exam 1-29   Unbroken lengths of busway shall be permitted to be extended through _____.

        A. dry walls
        B. hoistways
        C. hazardous locations
        D. damp locations

Timed Exam 1-30   Paper insulated conductors typically have what outer covering:

        A. Copper or alloy steel
        B. None
        C. Lead sheath
        D. Glass braid

Timed Exam 1-31 A means shall be provided in each metal box for the connection of an equipment grounding conductor. The means shall be permitted to be a _____ or equivalent.

    A. tapped hole
    B. toggle bolt
    C. cable clamp
    D. strap

Timed Exam 1-32 It shall be permissible to apply a demand factor of 75 percent to the nameplate rating load of _____ or more appliances fastened in place, other than electric ranges, clothes dryers, space-heating equipment, or air-conditioning equipment, that are served by the same feeder or service in a one-family, two-family, or multifamily dwelling.

    A. one
    B. two
    C. three
    D. four

Timed Exam 1-33 A device that provides a means for connecting intersystem bonding conductors for communications systems to the grounding electrode system.

    A. Grounding Electrode Conductor
    B. Intersystem Bonding Termination.
    C. Grounding Electrode
    D. Explosionproof Apparatus

Timed Exam 1-34 The minimum number of branch circuits shall be determined from the total _____ load and the size or rating of the circuits used.

    A. calculated
    B. area of
    C. amperage
    D. voltage

Timed Exam 1-35   The identification of terminals to which a grounded conductor is to be connected shall be substantially _____ in color and the identification of other terminals shall be of a readily distinguishable different color.

        A. green
        B. white
        C. gold
        D. black

Timed Exam 1-36   Where the opening to an outlet, junction, or switch point is less than _____ in. in any dimension, each conductor shall be long enough to extend at least _____ in. outside the opening. Exception ignored.

        A. 8, 3
        B. 6, 4
        C. 4, 3
        D. 6, 6

Timed Exam 1-37   Service-entrance conductors to buildings or enclosures shall be installed shall not be smaller than _____ AWG unless in multiconductor cable. Multiconductor cable shall not be smaller than _____ AWG.

        A. 8, 10
        B. 4, 6
        C. 6, 8
        D. 6, 4

Timed Exam 1-38   Circuits not exceeding 120 volts, nominal, between conductors shall be permitted to supply the following which of the following:

        A. The terminals of lampholders applied within their voltage ratings
        B. Cord-and-plug-connected or permanently connected utilization equipment
        C. The terminals of lampholders applied within their voltage ratings
        D. Auxiliary equipment of electric-discharge lamps
        E. All of the above

Timed Exam 1-39   Operation of equipment in excess of normal, full-load rating, or of a conductor in excess of rated ampacity that, when it persists for a sufficient length of time, would cause damage or dangerous overheating. A fault, such as a short circuit or ground fault, is not an overload.

       A. Ground Fault
       B. Arc Fault
       C. Overload
       D. Short Circuit

Timed Exam 1-40   What is the maximum number of conductors or fixture wires that can be used in electrical metallic tubing for No. 3 Type TW conductors in trade size 63 tubing.

       A. 8
       B. 10
       C. 15
       D. 20

Timed Exam 1-41   In circuits for control of irrigation and landscape lighting limited to not more than 30 volts and installed with type UF cable under minimum of 4-in. thick concrete exterior slab with no vehicular traffic and the slab extending not less than 6 in. beyond the underground installation what is the minimum required burial depth.

       A. 6
       B. 12
       C. 18
       D. 24

Timed Exam 1-42   Maximum Number of Disconnects, Two or three single-pole switches or breakers capable of individual operation shall be permitted on multiwire circuits, one pole for each ungrounded conductor, as one multipole disconnect, provided they are equipped with identified handle ties or a master handle to disconnect all ungrounded conductors with no more than _____ operations of the hand.

       A. six
       B. twelve
       C. four
       D. eight

Timed Exam 1-43   A complete lighting unit consisting of a light source such as a lamp or lamps, together with the parts designed to position the light source and connect it to the power supply. It may also include parts to protect the light source or the ballast or to distribute the light.

       A. Luminaire
       B. Lampholder
       C. Outline Lighting
       D. Lamp

Timed Exam 1-44   All storage or instantaneous-type water heaters shall be equipped with a temperature-limiting means in addition to its control thermostat to disconnect all _____.

       A. connected conductors
       B. ungrounded conductors
       C. water supply systems
       D. none of the above

Timed Exam 1-45   Electrical equipment and wiring and other electrically conductive material likely to become energized shall be installed in a manner that creates a low-impedance circuit facilitating the operation of the overcurrent device or ground detector for high-impedance grounded systems. It shall be capable of safely carrying the maximum ground-fault current likely to be imposed on it from any point on the wiring system where a ground fault may occur to the electrical supply source.

       A. GFI
       B. Bonding
       C. Effective Ground-Fault Current Path.
       D. grounded conductor

Timed Exam 1-46   The building or structure disconnecting means shall plainly indicate whether it is in the _____ position.

       A. hot or cold
       B. normal
       C. correct
       D. open or closed

Timed Exam 1-47 In kitchens, pantries, breakfast rooms, dining rooms, and similar areas of dwelling units, _____ small-appliance branch circuits shall serve all wall and floor receptacle outlets and all countertop outlets and receptacle outlets for refrigeration equipment. Exceptions ignored.

        A. one
        B. two
        C. two or more 20-ampere
        D. at least three

Timed Exam 1-48 A raceway consisting of grounded metal enclosure containing factory-mounted, bare or insulated conductors, which are usually copper or aluminum bars, rods, or tubes.

        A. Raceway
        B. Collector
        C. Busway
        D. Grounding Bus

Timed Exam 1-49 Where a feeder supplies branch circuits in which equipment grounding conductors are required, the feeder shall include or provide a(n) _____, in accordance with the provisions of 250.134, to which the equipment grounding conductors of the branch circuits shall be connected.

        A. water pipe
        B. chase connection
        C. grounded earth electrode
        D. equipment grounding conductor

Timed Exam 1-50 _____ covers shall be installed on all boxes, fittings, and similar enclosures to prevent accidental contact with energized parts or physical damage to parts or insulation.

        A. Suitable
        B. Metal
        C. Metal or Plastic
        D. Protective

Timed Exam 1-51   A box or conduit body shall not be required where a luminaire (fixture) is used as a _____ .

    A. junction box
    B. service panel
    C. grounding point
    D. raceway

Timed Exam 1-52   Normally non-current-carrying conductive materials enclosing electrical conductors or equipment, or forming part of such equipment, shall be _____ so as to limit the voltage to ground on these materials.

    A. marked with a warning label
    B. insulated
    C. connected to earth
    D. ground fault protected

Timed Exam 1-53   In walls of concrete, tile, or other noncombustible material, cabinets shall be installed so that the front edge of the cabinet is not set back of the finished surface more than _____ in.

    A. 1/8
    B. 1/4
    C. 3/8
    D. 1/2

Timed Exam 1-54   Where conductors carrying alternating current are installed in ferrous metal enclosures or ferrous metal raceways, they shall be arranged so as to avoid heating the surrounding ferrous metal by induction by _____ . Exceptions ignored.

    A. installing conductor spacers
    B. coiling like phase conductors
    C. grouping all phase conductors together
    D. none of the above

Timed Exam 1-55   In dwelling units, at least one _____ outlet shall be installed in every habitable room and bathroom.   Exceptions ignored.

        A. appliance
        B. wall switch-controlled lighting
        C. hair dryer
        D. 20 amp.

Timed Exam 1-56   Where a raceway enters a building or structure from an underground distribution system, it shall be _____ in accordance with 300.5(G). Spare or unused raceways shall also be _____.

        A. labeled
        B. sealed
        C. listed
        D. insulated

Timed Exam 1-57   Branch circuits larger than 50 amperes shall supply only _____ outlet loads.

        A. non-lighting
        B. lighting
        C. ranges
        D. cooking equipment

Timed Exam 1-58   What is the maximum number of conductors or fixture wires that can be used in electrical metallic tubing for No. 14 Type THHW conductors in trade size 27 tubing.

        A. 6
        B. 10
        C. 16
        D. 28

Timed Exam 1-59 The copper sheath of mineral-insulated, metal-sheathed cable Type MI may be used as the equipment grounding conductor.

        A. This is true
        B. This is false

Timed Exam 1-60 Switches and circuit breakers used as disconnecting means shall be of the _____ type.

        A. toggle
        B. T
        C. indicating
        D. grounded disconnect conductor

Timed Exam 1-61 Taps from bare conductors shall leave the gutter _____ their terminal connections, and conductors shall not be brought in contact with uninsulated current-carrying parts of different potential.

        A. opposite
        B. parallel to
        C. within 3 inches of
        D. bare from

Timed Exam 1-62 In dwelling Units, all 125-volt, single-phase, 15- and 20-ampere receptacles installed in _____ shall have ground-fault circuit-interrupter protection for personnel. Exceptions ignored.

        A. Garages, and also accessory buildings that have a floor located at or below grade level not intended as habitable rooms and limited to storage areas, work areas, and areas of similar use.
        B. Bedrooms
        C. Bathrooms
        D. Both A and C

Timed Exam 1-63   A building or structure electrical disconnecting means can be _____.

      A. located where ever practicle
      B. protected from public tampering by locks or disguises
      C. electrically operated by a readily accessible, remote-control device in a separate building or structure.
      D. all of the above

Timed Exam 1-64   Which of the following methods is not approved for the installation of service-entrance conductors.

      A. Electrical metallic tubing
      B. Busways
      C. Type NM cable
      D. Type MC cable

Timed Exam 1-65   The equipment grounding conductor can be the casing enclosing the circuit conductors and may be in the form of any of the following except :_____

      A. RMC
      B. RNC
      C. IMT
      D. EMT

Timed Exam 1-66   Direct-buried conductors or cables shall be permitted to be spliced or tapped _____.

      A. without the use of splice boxes
      B. only with the use of splice boxes
      C. only with the use of waterproof splice boxes
      D. and tapped with a minimum of 1/2" of friction tape

Timed Exam 1-67   A applied to circuit breakers, a qualifying term indicating that there is purposely introduced a delay in the tripping action of the circuit breaker, which delay decreases as the magnitude of the current increases.

       A. Delayed Trip
       B. Slow Blow
       C. Time Protected
       D. Inverse Time

Timed Exam 1-68   Bends shall be made so that the tubing is not damaged and the internal diameter of the tubing is not effectively _____.

       A. increased
       B. reduced
       C. lengthened
       D. non of the above

Timed Exam 1-69   Switches or circuit breakers shall not disconnect the _____ conductor of a circuit. Exception ignored.

       A. ungrounded
       B. energized
       C. grounded
       D. hot

Timed Exam 1-70   Electrical equipment that depends on the natural circulation of air and _____ for cooling of exposed surfaces shall be installed so that room airflow over such surfaces is not prevented by walls or by adjacent installed equipment.

       A. independent fans
       B. heat sinks
       C. convection principles
       D. direct contact of cooling fins

Timed Exam 1-71   The dwelling has a floor area of 1500 sq. ft. exclusive of an unfinished cellar not adaptable for future use, unfinished attic, and open porches. It has two 20-A small appliance circuits, one 20-A laundry circuit, two 4-kW wall-mounted ovens, one 5.1-kW counter-mounted cooking unit, a 4.5-kW water heater, a 1.2-kW dishwasher, a 5-kW combination clothes washer and dryer, six 7-A, 230-V room air-conditioning units, and a 1.5-kW, 230 V permanently installed bathroom space heater. Assume wall-mounted ovens, counter-mounted cooking unit, water heater, dishwasher, and combination clothes washer and dryer kW ratings equivalent to kVA. Assuming the dwelling is feed by a 120/240-V, 3-wire, single-phase service what is the total calculated load in VA and the minimum required service rating required.

        A. 19,200 VA, 100 A
        B. 25,700 VA, 110 A
        C. 29,200 VA, 122 A
        D. 39,200 VA, 175 A

Timed Exam 1-72   What is the maximum percent of FMC conduit total interior area that can be filled if more that 2 conductors are used in the tube.

        A. 53
        B. 31
        C. 40
        D. 60

Timed Exam 1-73   _____ used in the construction of a luminaire (fixture) box.

        A. Screws, crimps or solder may be
        B. Only silver solder may be
        C. No solder shall be
        D. none of the above

Timed Exam 1-74   A multifamily dwelling contains 11 units. What is the demand factor to be used when calculating the service load.

        A. 50
        B. 45
        C. 42
        D. 37

Timed Exam 1-75   The general calculated load shall be not less than _____ percent of the first 10 kVA plus _____ percent of the remainder of the loads.

        A. 100, 125
        B. 80, 100
        C. 100, 40
        D. 90, 50

Timed Exam 1-76   An insulated grounded conductor of 6 AWG or smaller shall be identified by _____.

        A. A continuous white outer finish.
        B. A continuous gray outer finish.
        C. Three continuous white or gray stripes along the conductor's entire length on other than green insulation.
        D. Wires that have their outer covering finished to show a white or gray color but have colored tracer threads in the braid identifying the source of manufacture shall be considered as meeting the provisions of this section.
        E. Any of the above and more options are available

Timed Exam 1-77   A reliable conductor to ensure the required electrical conductivity between metal parts required to be electrically connected.

        A. Bonding Jumper
        B. Connection
        C. Soldering
        D. Welding

Timed Exam 1-78   The calculated lighting load for an office building is 24,500 VA. What is the maximum sq ft that the office could have if this were the lighting load.

        A. 4500
        B. 5600
        C. 7000
        D. 8300

Timed Exam 1-79   Each commercial building and each commercial occupancy accessible to pedestrians shall be provided with at least one outlet in an accessible location at each _____ to each tenant space for sign or outline lighting system use.

        A. exit
        B. store front
        C. window
        D. entrance

Timed Exam 1-80   Receptacles that provide power for water-pump motors or for other loads directly related to the circulation and sanitation system shall be located at least 10 ft from the inside walls of the pool, or not less than 6 ft from the inside walls of the pool if they meet which of the following conditions.

        A. Consist of single receptacles
        B. Employ a locking configuration
        C. Are of the grounding type
        D. Have GFCI protection
        E. all of the above

Timed Exam 2   31/2 Hours to Complete

Timed Exam 2-1   A dwelling has a floor area of 1500 sq. ft, with unfinished cellar not adaptable for future use, unfinished attic, and open porches. Appliances are a 12-kW range and a 5.5-kW, 240-V dryer. Assume range and dryer kW ratings equivalent to kVA ratings in accordance with 220.54 and 220.55. What is the minimum number of branch general light circuits required.

         A. Three 15-A, 2-wire or two 20-A, 2-wire circuits
         B. Five 15-A, 2-wire or three 20-A, 2-wire circuits
         C. Two 15-A, 2-wire and two 20-A, 2-wire circuits
         D. Four 15-A, 2-wire or Three 20-A, 2-wire circuits

Timed Exam 2-2   What is the maximum allowable ampacity for No. 2/0 copper conductors type THHN, 167 F rated insulation, installed in an ambient temperature of 30 C. There are 2 current carrying conductors in the cable.

         A. 135
         B. 150
         C. 115
         D. 195

Timed Exam 2-3   In a service installation for a dwelling what is the maximum service feeder rating for No. 4 copper service conductors in amperes.

         A. 100
         B. 125
         C. 200
         D. 225

Timed Exam 2-4   In hospitals the lighting demand factor of the _____ VA is 20 Percent.

         A. first 3000
         B. Remainder over 50,000
         C. VA Load from 3001 to 120,000
         D. all lighting

Timed Exam 2-5   In general receptacles shall be not less than _____ ft from the inside walls of a pool.

        A. 5
        B. 6
        C. 15
        D. 25

Timed Exam 2-6   Overhead Service, where not in excess of 1000 volts, nominal, shall follow minimum clearances measured from _____.

        A. the top of the mast
        B. final grade
        C. the guy wire anchor
        D. the foundation wall

Timed Exam 2-7   Conductors other than _____ shall not be installed in the same service raceway or service cable. Exceptions ignored.

        A. copper
        B. service conductors
        C. ungrounded
        D. SE or USE

Timed Exam 2-8   Two-wire dc circuits and ac circuits of two or more _____ conductors shall be permitted to be tapped from the _____ conductors of circuits that have a grounded neutral conductor.

        A. ungrounded
        B. grounding
        C. grounded
        D. hot leg

Timed Exam 2-9   ENT shall not be used for the _____ of luminaries and other equipment.

        A. connection
        B. wiring method
        C. support
        D. conductors

Timed Exam 2-10   FMC is not permitted in _____.

    A. In storage battery rooms
    B. Concealed spaces
    C. Exposed locations
    D. Interior walls

Timed Exam 2-11   For overhead conductors near a swimming pool for Insulated cables, 0-750 Volts to ground, supported on and cabled together with an effectively grounded bare messenger or effectively grounded neutral conductor, what is the clearance in any direction to the observation stand, tower, or diving platform.

    A. 22.5 feet
    B. 25 feet
    C. 14.5 feet
    D. 17 feet

Timed Exam 2-12   A unit load of not less than _____ VA per square foot is required for general lighting in Hospital occupancies.

    A. 1
    B. 2
    C. 3
    D. 4

Timed Exam 2-13   A ground ring encircling the building or structure, in direct contact with the earth, consisting of at least 20 ft of bare copper conductor not smaller than _____ AWG shall be permitted to be used as a grounding electrode.

    A. 2/0
    B. 1
    C. 2
    D. 4

Timed Exam 2-14   Where the building or structure disconnecting means does not disconnect the _____ conductor from the _____ conductors in the building or structure wiring, other means shall be provided for this purpose at the location of disconnecting means. A terminal or bus to which all grounded conductors can be attached by means of pressure connectors shall be permitted for this purpose.

      A. grounded
      B. ungrounded
      C. grounding
      D. uninsulated

Timed Exam 2-15   A dwelling has a floor area of 1500 sq. ft, with unfinished cellar not adaptable for future use, unfinished attic, and open porches. Appliances are a 12-kW range and a 5.5-kW, 240-V dryer. Assume range and dryer kW ratings equivalent to kVA ratings in accordance with 220.54 and 220.55. What is the total net calculated load for the dwelling in VA units.

      A. 15,600
      B. 30,500
      C. 18,600
      D. 22,600

Timed Exam 2-16   In feeders over 600 Volts Supplying Transformers and Utilization Equipment, the ampacity of feeders supplying a combination of transformers and utilization equipment shall not be less than the sum of the nameplate ratings of the transformers and 125 percent of the designed potential load of the utilization equipment that will be operated _____.

      A. independently
      B. non continuously
      C. simultaneously
      D. continuously

Timed Exam 2-17   The maximum current, in amperes, that a conductor can carry continuously under the conditions of use without exceeding its temperature rating.

        A. Ampacity
        B. Voltage
        C. Power Factor
        D. Capacitance

Timed Exam 2-18   The complete electrical system shall be performance tested when _____.

        A. before occupancy
        B. first installed on-site
        C. at 3 separate stages during construction.
        D. on a monthly basis

Timed Exam 2-19   What is the minimum bending radius in inches for 3/4" EMT using a Full Shoe bender.

        A. 3.5
        B. 4.5
        C. 5.5
        D. 7.25

Timed Exam 2-20   The minimum cover requirements for underground wiring 0 to 600 Volts, in Type UF cable under streets, highways, roads, alleys, driveways, and parking lots is _____ inches.

        A. 6
        B. 12
        C. 18
        D. 24

Timed Exam 2-21   Throughout the Code, the voltage considered shall be that at which the circuit operates, and the voltage rating of electrical equipment shall not be less than the _____ of a circuit to which it is connected.

        A. under current protection rating
        B. over current protection rating
        C. nominal voltage
        D. stable effective voltage

Timed Exam 2-22   The NEC (National Electrical Code) provided that the number of wires and circuits confined in a single enclosure be _____. Limiting the number of circuits in a single enclosure minimizes the effects from a short circuit or ground fault in one circuit.

        A. varyingly restricted
        B. restricted to 3 circuits
        C. restricted to 1 circuit
        D. unrestricted

Timed Exam 2-23   Where cable is used, each cable shall be _____ to the cabinet, cutout box, or meter socket enclosure. Exceptions ignored

        A. secured
        B. clamped
        C. bonded
        D. affixed

Timed Exam 2-24   Receptacles and cord connectors shall be rated not less than _____ amperes, 125 volts, or 250 volts, and shall be of a type not suitable for use as lampholders.

        A. 12
        B. 15
        C. 20
        D. 25

Timed Exam 2-25   Where a submersible pump is used in a metal well casing, the well casing shall be _____ to the pump circuit equipment grounding conductor.

        A. tied
        B. clamped
        C. bonded
        D. connected

Timed Exam 2-26   Cable tray systems use to support service-entrance conductors shall contain only _____ conductors. Exception ignored.

        A. service-entrance
        B. Type MC cable
        C. branch feeder
        D. insulated

Timed Exam 2-27   Switching devices shall be located at least _____ ft horizontally from the inside walls of a pool unless separated from a pool by a solid fence, wall, or other permanent barrier. Alternatively, a switch that is listed as being acceptable for use within this distance shall be permitted.

        A. 5
        B. 10
        C. 15
        D. 20

Timed Exam 2-28   The common point on a wye-connection in a polyphase system or midpoint on a single-phase, 3-wire system, or midpoint of a single-phase portion of a 3-phase delta system, or a midpoint of a 3-wire, direct-current system.

        A. Plenum
        B. Switching Device
        C. Service Point
        D. Neutral Point.

Timed Exam 2-29   Fuses and circuit breakers shall be permitted to be connected in parallel where they are _____ and listed as a unit. Individual fuses, circuit breakers, or combinations thereof shall not otherwise be connected in parallel.

        A. rated equally for amperage and time delay
        B. matched pairs
        C. factory assembled in parallel
        D. of the same type and rating

Timed Exam 2-30   Continuous Duty

    A. Operation at a substantially constant load for an indefinitely long time.
    B. Operation at a substantially constant load for an 2 or more hours.
    C. Operation at a substantially constant voltage for an indefinitely long time.
    D. Operation at a substantially constant load for the normal operating requirements of the device.

Timed Exam 2-31   In walls constructed of wood or other combustible material, cabinets shall be flush with the finished surface _____.

    A. or project therefrom
    B. or recessed not more than 1/4 inch
    C. or set off of the surface 1/4 inch.
    D. placed on a metal plate

Timed Exam 2-32   Fuses and circuit breakers shall be _____ so that persons will not be burned or otherwise injured by their operation.

    A. located or shielded
    B. insulated
    C. non-accessible
    D. non of the above

Timed Exam 2-33   A hand-operable circuit breaker equipped with a lever or handle, or a power-operated circuit breaker capable of being opened by hand in the event of a power failure, shall be permitted to serve as a switch if it has _____.

    A. a plastic insulated handle
    B. the required number of poles
    C. not been mounted in a panelboard
    D. has been mounted in a panelboard

Timed Exam 2-34   In dwelling units, at least one receptacle outlet shall be installed in bathrooms within _____ of the outside edge of each basin and the receptacle outlet shall be located on a wall or partition that is adjacent to the basin or basin countertop, located on the countertop, or installed on the side or face of the basin cabinet not more than 300 mm (12 in.) below the countertop. Receptacle outlet assemblies listed for the application shall be permitted to be installed in the countertop

        A. 3 ft
        B. 6 ft
        C. 4 ft
        D. 18 in.

Timed Exam 2-35   Each unit length of a heating cable a blue lead wire shall indicates what voltage requirement for that element.

        A. 120
        B. 208
        C. 240
        D. 277

Timed Exam 2-36   At least _____ in. of free conductor, measured from the point in the box where it emerges from its raceway or cable sheath, shall be left at each outlet, junction, and switch point for splices or the connection of luminaries (fixtures) or devices. Exception ignored.

        A. 3
        B. 4
        C. 6
        D. 8

Timed Exam 2-37   Bonding jumpers shall be of copper or other corrosion-resistant material, a bonding jumper shall be a _____ or similar suitable conductor.

        A. wire
        B. bus
        C. screw
        D. any of the above

Timed Exam 2-38   Continuity of the grounding path or the bonding connection to interior piping shall not rely on _____.

       A. water meters or filtering devices and similar equipment
       B. grounding jumpers
       C. patch conductors and connectors
       D. pressure connectors

Timed Exam 2-39   Overhead spans of open conductors and open multiconductor cables of not over 600 volts, nominal, shall have a clearance of not less than _____ ft over residential property and driveways, and those commercial areas not subject to truck traffic where the voltage does not exceed 300 volts to ground.

       A. 10
       B. 12
       C. 14
       D. 18

Timed Exam 2-40   Externally Operable

       A. Equipment enclosed in a case that is capable of withstanding an explosion of a specified gas or vapor that may occur within it and of preventing the ignition of a specified gas or vapor surrounding the enclosure by sparks, flashes, or explosion of the gas or vapor within, and that operates at such an external temperature that a surrounding flammable atmosphere will not be ignited thereby.
       B. Capable of being operated without exposing the operator to contact with live parts.
       C. Capable of being inadvertently touched or approached nearer than a safe distance by a person. It is applied to parts that are not suitably guarded, isolated, or insulated.
       D. Incapable of being operated without exposing the operator to contact with live parts.

Timed Exam 2-41   In kitchens where the receptacles are installed to serve the countertop surfaces outlets within _____ shall be GFI protected.

        A. 3 ft. of the sink
        B. 6 ft. of the sink
        C. 24 ft. of the sink
        D. all these outlets require GFI protection

Timed Exam 2-42   The grounding electrode conductor material selected shall be resistant to any _____ condition existing at the installation or shall be protected against corrosion.

        A. moisture
        B. high temperature
        C. corrosive
        D. below freezing

Timed Exam 2-43   Equipment that has an open-circuit voltage exceeding _____ volts shall not be installed in or on dwelling occupancies.

        A. 208
        B. 240
        C. 1000
        D. 1200

Timed Exam 2-44   Electrical installations in hollow spaces, vertical shafts, and ventilation or air-handling ducts shall be made so that the possible spread of fire or products of combustion will not be substantially increased. Openings around electrical penetrations into or through fire-resistant-rated walls, partitions, floors, or ceilings shall be _____ using approved methods to maintain the fire resistance rating.

        A. blocked
        B. sealed
        C. caulked or taped
        D. firestopped

Timed Exam 2-45   A 15- or 20-ampere branch circuit shall be permitted to supply lighting units or other utilization equipment, or a combination of both. Exceptions ignored.

        A. This is True
        B. This is False

Timed Exam 2-46   In swimming pool installations, where none of the required bonded parts is in direct connection with the pool water, the pool water shall be in direct contact with an approved corrosion-resistant conductive surface that exposes not less than _____ of surface area to the pool water at all times. The conductive surface shall be located where it is not exposed to physical damage or dislodgement during usual pool ac- tivities, and

        A. 9 sq. in.
        B. 81 sq. in.
        C. 100 sq. in.
        D. 300 sq. in.

Timed Exam 2-47   Devices such as pressure terminal or pressure splicing connectors and soldering lugs shall be identified for the _____ and shall be properly installed and used.

        A. material of the conductor
        B. location of installation
        C. size of conductors
        D. proper spacing

Timed Exam 2-48   In the installation and uses of electric wiring and equipment in ducts, plenums, and other air-handling spaces, _____ shall be installed in ducts or shafts containing only such ducts, used for vapor removal or for ventilation of commercial-type cooking equipment.

        A. no wiring systems of any type
        B. only rigid metallic conduit
        C. no flexible type conduit
        D. none of the above

Timed Exam 2-49   In a one family dwelling and each unit of a two-family dwelling that is at grade level, at least one receptacle outlet readily accessible from grade and not more than _____ above grade shall be installed at the front and back of the dwelling.

      A. 6 1/2 ft
      B. 78 in.
      C. 6.5 feet
      D. all of the above

Timed Exam 2-50   A premises wiring system supplied by a grounded ac service shall have a _____ connected to the grounded service conductor, at each service. Exception ignored.

      A. lightening arrester
      B. circuit breaker
      C. grounding electrode conductor
      D. bonding jumper

Timed Exam 2-51   Conductor sizes are expressed in American Wire Gage (AWG) or in _____.

      A. sq. inches
      B. standard metric sizes
      C. standard English system sizes
      D. circular mils

Timed Exam 2-52   _____ of conductors on buildings, structures, or poles shall be as provided for services in 230.50.

      A. Proper spacing
      B. Insulating values
      C. Mechanical protection
      D. Minimum height clearances

Timed Exam 2-53   In a multiwire branch circuit all conductors shall originate from the same _____ or similar distribution equipment.

        A. location
        B. panelboard
        C. junction box
        D. meter

Timed Exam 2-54   The branch-circuit conductors supplying one or more units of a information technology equipment shall have an ampacity not less than _____ percent of the total connected load.

        A. 80
        B. 100
        C. 125
        D. 150

Timed Exam 2-55   The purpose of the National Electrical Code is the _____ of persons and property from hazards arising from the use of electricity.

        A. Economical protection
        B. Protection
        C. Adequate Safety
        D. Practical Safeguarding.

Timed Exam 2-56   The lightning protection system ground terminals shall be bonded to the building or structure _____.

        A. lighting rod electrode
        B. frame
        C. foundation rebar
        D. grounding electrode system

Timed Exam 2-57   Where more than one ground rod is used for grounding the electrodes, each electrode of one grounding system (including that used for strike termination devices) shall not be less than ____ ft from any other electrode of another grounding system.

        A. 4
        B. 6
        C. 8
        D. 10

Timed Exam 2-58   Sheet metal auxiliary gutters shall be supported and secured throughout their entire length at intervals not exceeding _____ ft.

     A. 3
     B. 5
     C. 6
     D. 8

Timed Exam 2-59   Circuit conductors shall be permitted to be installed in raceways; in cable trays; as metal-clad cable, as bare wire, cable Thpe MC: as bare wire, cable, and busbars; or as Type MV cables or conductors as provided in 300.37, 300.39, 300.40, and 300.50. Bare live conductors shall _____.

     A. not be allowed
     B. conform with 490.24.
     C. be installed in PVC RNC.
     D. be marked as live.

Timed Exam 2-60   What is the maximum ampere rating of a insulated single copper conductor in air with a conductor temperature of 90 C. in an ambient temperature of 40 C , the conductor size is No. 4 and the voltage is 2500 Volts.

     A. 100
     B. 110
     C. 115
     D. 145

Timed Exam 2-61   Receptacles rated 20 amperes or less and designed for the direct connection of aluminum conductors shall be marked _____.

     A. CO/ALR
     B. Aluminum
     C. Any Metal
     D. Bi-Metal

Timed Exam 2-62   On overhead service-drop individual conductors shall be insulated or covered, however, the _____ conductor of a multiconductor cable shall be permitted to be bare.

        A. grounding
        B. isolated hot leg
        C. ungrounded
        D. grounded

Timed Exam 2-63   Where a change occurs in the size of the ungrounded conductor, a similar change shall be permitted to be made in the size of the _____ conductor.

        A. grounded
        B. grounding
        C. bonding
        D. neutral

Timed Exam 2-64   For dwelling units, attached garages, and detached garages with electric power, at least one _____ shall be installed to provide illumination on the exterior side of outdoor entrances or exits with grade level access.

        A. Flood light
        B. Motion detector
        C. Security camera
        D. wall switch controlled lighting outlet

Timed Exam 2-65   Where a flexible cord is used to supply a room air conditioner, the length of such cord shall not exceed _____ ft for a nominal, 120-volt rating _____ ft for a nominal, 208- or 240-volt rating.

        A. 8, 10
        B. 6, 4
        C. 10, 6
        D. 6, 3

Timed Exam 2-66   Service conductors installed as open conductors or multiconductor cable without an overall outer jacket shall have a clearance of not less than _____ ft from windows that are designed to be opened, doors, porches, balconies, ladders, stairs, fire escapes, or similar locations. Exceptions ignored.

   A. 3
   B. 5
   C. 8
   D. 10

Timed Exam 2-67   Where an ac system is connected to a grounding electrode in or at a building or structure, _____ electrode(s) shall be used to ground conductor enclosures and equipment in or on that building or structure.

   A. separate
   B. independent
   C. non-bonded
   D. the same

Timed Exam 2-68   Where can additional grounding and bonding requirements for natural and artificially made bodies of water be found in the NEC.

   A. Article 517
   B. Article 230
   C. Article 690
   D. Article 682

Timed Exam 2-69   Overcurrent protection shall be provided in each ungrounded circuit conductor and shall be located _____. Exceptions ignored.

   A. at the point where the conductors receive their supply
   B. at each end of a conductor
   C. at a convent service point in the circuit
   D. on the load end of the circuit

Timed Exam 2-70   Wiring located within the cavity of a fire-rated floor-ceiling or roof-ceiling assembly shall not be secured to, or supported by _____. Exceptions ignored.

        A. metal cross members
        B. the ceiling assembly, including the ceiling support wires
        C. joists, beams, or columns
        D. cantilevered floor joists

Timed Exam 2-71   A raceway of circular cross section made of helically wound, formed, interlocked metal strip.

        A. Electrical Metallic Tubing (EMT)
        B. Rigid Metal Conduit (RMC)
        C. Flexible Metal Conduit (FMC)
        D. Rigid Non-metallic Conduit (RNC)

Timed Exam 2-72   Other than for motor overload protection, no overcurrent device shall be connected in series with any conductor that is intentionally grounded, the overcurrent device opens all conductors of the circuit, including the grounded conductor, and is designed so that no pole can operate _____.

        A. simultaneously
        B. above the open limit
        C. until the circuit breaker for each is closed
        D. independently

Timed Exam 2-73   Where rock bottom is encountered in installing a grounding rod electrode the electrode shall be driven at an oblique angle not to exceed _____ degrees from the vertical or, where rock bottom is encountered at an angle up to 45 degrees.

        A. 30
        B. 45
        C. 60
        D. 75

Timed Exam 2-74 In both exposed and concealed locations, where a cable or raceway type wiring method is installed through bored holes in joists, rafters, or wood members, holes shall be bored so that the edge of the hole is not less than _____ in. from the nearest edge of the wood member. Where this distance cannot be maintained, the cable or raceway shall be protected from penetration by screws or nails by a steel plate or bushing, at least _____ in. thick, and of appropriate length and width installed to cover the area of the wiring. Exceptions ignored.

        A. 1.5, .125
        B. 1.625, .20
        C. 1.25, .0625
        D. 1.75, .25

Timed Exam 2-75 A receptacle outlet shall be installed at each wall countertop space that is _____ or wider. Receptacle outlets shall be installed so that no point along the wall line is more than _____ measured horizontally from a receptacle outlet in that space. Exceptions ignored.

        A. 24 in., 48 in.
        B. 18 in., 36 in.
        C. 12 in., 24 in.
        D. 2 ft., 4 ft.

Timed Exam 2-76 A device designed to open and close a circuit by nonautomatic means and to open the circuit automatically on a predetermined overcurrent without damage to itself when properly applied within its rating.

        A. Fuse
        B. Circuit Breaker
        C. Cartridge Fuse
        D. Fusible Link

Timed Exam 2-77 Lighting track may be installed in the following locations.

        A. In storage battery rooms
        B. In wet or damp locations
        C. Where likely to be subjected to physical damage
        D. Where concealed

E. none of the above

Timed Exam 2-78   A unit load of not less than _____ per square foot is required for general lighting in Church occupancies.

       A. 1 VA
       B. 2 VA
       C. 3 VA
       D. 4 VA

Timed Exam 2-79   Where the number of current-carrying conductors in a raceway or cable exceeds _____, or where single conductors or multiconductor cables are installed without maintaining spacing for a continuous length than _____ in. and are not installed in raceways, the allowable ampacity of each conductor shall be reduced by a derating adjustment factor. Exceptions ignored.

       A. three, 24
       B. six, 48
       C. twelve, 24
       D. fifteen, 48

Timed Exam 2-80   A reliable conductor to ensure the required electrical conductivity between metal parts required to be electrically connected.

       A. Branch Circuit
       B. Bonding Jumper
       C. Copper Wire
       D. Ground Wire

Timed Exam 3   3 1/2 Hours to Complete

Timed Exam 3-1   A wall space in a dwelling unit for purposes of outlet spacing shall be considered any unbroken any space _____ ft or more in width including space measured around corners and unbroken along the floor line by doorways, fireplaces, and similar openings.

    A. 1
    B. 2
    C. 3
    D. 4

Timed Exam 3-2   Liquidtight Flexible Nonmetallic Conduit (LFNC) is permitted to be used _____. Exceptions Ignored.

    A. Where protection of the contained conductors is required from vapors, liquids, or solids
    B. In lengths longer than 100 ft,
    C. Where the operating voltage of the contained conductors is in excess of 600 volts, nominal,
    D. In any conditions of extreme cold or heat.

Timed Exam 3-3   An insulated conductor that is intended for use as a grounded conductor, where contained within a flexible cord, shall be identified by a _____ outer finish or by methods permitted by 400.22.

    A. white or gray
    B. green
    C. bare
    D. varnished

Timed Exam 3-4   High Density Polyethylene (HDPE) Conduit a nonmetallic raceway of circular cross section, with associated couplings, connectors, and fittings for the installation of electrical conductors is not allowed to be installed _____.

    A. In discrete lengths or in continuous lengths from a reel
    B. In cinder fill
    C. In locations subject to severe corrosive influences as covered in 300.6 and where subject to chemicals for which the conduit is listed
    D. Within a building

Timed Exam 3-5   Conductors normally used to carry current shall be of _____ unless otherwise provided in this Code and where the conductor material is not specified, the material and the sizes given in the Code shall apply to copper conductors.

        A. copper
        B. aluminum
        C. copper clad
        D. copper alloy

Timed Exam 3-6   For receptacles in damp or wet locations, an installation suitable for wet locations shall also be considered suitable for _____ locations.

        A. underwater
        B. damp
        C. exposed
        D. corrosive vapor

Timed Exam 3-7   A switch intended for use in general distribution and branch circuits. It is rated in amperes, and it is capable of interrupting its rated current at its rated voltage.

        A. Isolation Switch
        B. General-Use Switch
        C. Toggle Switch
        D. Breaker Switch

Timed Exam 3-8   Type NM cables shall be durably marked on the surface, the AWG size or circular mil area shall be repeated at intervals not exceeding _____ in., and all other markings shall be repeated at intervals not exceeding _____ in..

        A. 12, 24
        B. 18, 48
        C. 24, 40
        D. 40, 24

Timed Exam 3-9   On a 4-wire, delta-connected system where the midpoint of one phase winding is grounded, only the conductor or busbar having the higher phase voltage to ground shall be durably and permanently marked by an outer finish that is _____ in color or by other effective means.

    A. Red
    B. Yellow
    C. Orange
    D. Black or Red

Timed Exam 3-10   For each yoke or strap containing one or more devices or equipment, a _____ volume allowance shall be made for each yoke or strap based on the largest conductor connected to a device(s) or equipment supported by that yoke or strap.

    A. single
    B. double
    C. zero
    D. multiple

Timed Exam 3-11   Where necessary to prevent tampering, an automatic overcurrent device that protects service conductors supplying only a specific load, such as a water heater, shall be permitted to be _____ where located so as to be accessible.

    A. locked or sealed
    B. openly accessible only
    C. hidden from view
    D. not labeled

Timed Exam 3-12   An outdoor branch circuit has conductors of 600 V and is strung between poles over a water area not suitable for boating. The conductors are a messenger wire supported twisted cable unit and if 15 ft above the water. You have inspected the installation and come to the following conclusion.

    A. The installation meets Code requirements.
    B. The installation is in violation of Code requirements because the height requirement is not met.
    C. The installation is in violation of Code requirements because the voltage requirement is not met.
    D. none of the above

Timed Exam 3-13   For attics and underfloor spaces containing equipment requiring servicing, such as heating, air-conditioning, and refrigeration equipment, at least one_____ containing a switch or controlled by a wall switch shall be installed in such spaces. At least one point of control shall be at the usual point of entry to these spaces.

        A. lighting outlet
        B. power control
        C. junction box
        D. panelboard

Timed Exam 3-14   "Completed wiring installations shall be free from short circuits and from _____ other than as required or permitted in Article 250.

Completed wiring installations shall be free from short circuits, _____, or any connections to ground other than as required or permitted elsewhere in this Code."

        A. switches
        B. ground faults
        C. fuses
        D. teminals

Timed Exam 3-15   THHN insulation has what which characteristic trade name.

        A. Heat-resistant thermoplastic
        B. Moisture-resistant thermoplastic
        C. Moisture- and heat-resistant thermoplastic
        D. Moisture, heat, and oil resistant thermoplastic

Timed Exam 3-16   A unit load of not less than _____ VA per square foot is required for general lighting in dwelling units.

        A. 1
        B. 2
        C. 3
        D. 4

Timed Exam 3-17   This Code does not covers the installation of electrical conductors, equipment, and raceways; signaling and communications conductors, equipment, and raceways; and optical fiber cables and raceways for _____.

    A. Installations of conductors and equipment that connect to the supply of electricity.
    B. Yards, lots, parking lots, carnivals, and industrial substations.
    C. Installations of communications equipment under the exclusive control of communications utilities located outdoors or in building spaces used exclusively for such installations.
    D. Public and private premises, including buildings, structures, mobile homes, recreational vehicles, and floating buildings.

Timed Exam 3-18   Metric designator 41 is the same as trade size _____ inch conduit.

    A. 1/2
    B. 1
    C. 1 1/2
    D. 2

Timed Exam 3-19   Service conductors passing over a roof shall be securely supported by substantial structures. Where practicable, such supports shall be_____.

    A. independent of the building
    B. coated with a non-conductive coating
    C. waterproof
    D. part of the buildings structure

Timed Exam 3-20   In dwelling units, where a _____ is installed in an island or peninsular countertop and the width of the countertop behind the range, counter-mounted cooking unit, or sink is less than 300 mm (12 in.), the _____ is considered to divide the countertop space into two separate countertop spaces.

    A. wall outlets
    B. wall switches
    C. pictures
    D. range, counter-mounted cooking unit, or sink

Timed Exam 3-21   In a 40 amp range circuit the ampere rating of a range receptacle shall be permitted to be _____ amps.

       A. 40
       B. 50
       C. 40 or 50
       D. all of the above

Timed Exam 3-22   Service disconnecting means shall not be installed in _____.

       A. hallways
       B. utility rooms
       C. bathrooms
       D. basements

Timed Exam 3-23   Where a service mast is used for the support of service-drop conductors, it shall be of adequate strength or be supported by _____ to withstand safely the strain imposed by the service drop.

       A. lateral cables
       B. angle ties
       C. rigid metallic tubing
       D. braces or guys

Timed Exam 3-24   A dwelling has a floor area of 2000 sq. ft exclusive of an unfinished cellar not adaptable for future use, unfinished attic, and open porches. It has a 12-kW range, a 4.5-kW water heater, a 1.2-kW dishwasher, a 5-kW clothes dryer, and a 2 1/2-ton (24-A) heat pump with 15 kW of backup heat. What is the total heating and cooling load in VA units.

       A. 5,760
       B. 15,000
       C. 20,760
       D. 15,510

Timed Exam 3-25   The point of connection between the facilities of the serving utility and the premises wiring.

        A. Meter Base
        B. Meter Socket
        C. Power Entrance
        D. Service Point

Timed Exam 3-26   Raceways on exteriors of buildings or other structures shall be arranged to drain and shall be _____ in wet locations.

        A. waterproof
        B. raintight
        C. rainproof
        D. suitable for use

Timed Exam 3-27   Bonding Conductor or Jumper

        A. Performs a function without the necessity of human intervention
        B. A generic term for a group of nonflammable synthetic chlorinated hydrocarbons used as electrical insulating media.
        C. The circuit conductors between the final overcurrent device protecting the circuit and the outlet(s).
        D. A reliable conductor to ensure the required electrical conductivity between metal parts required to be electrically connected.

Timed Exam 3-28   IMC shall be securely fastened within ____ ft of each outlet box, junction box, device box, cabinet, conduit body, or other conduit termination. Fastening shall be permitted to be increased to a distance of ____ ft where structural members do not readily permit fastening at the standard minimum distance. Exceptions ignored.

        A. 5, 10
        B. 4, 10
        C. 4, 8
        D. 3, 5

Timed Exam 3-29   Where connected to a branch circuit having a rating in excess of 20 amperes, lampholders shall be of _____ and such lampholder shall have a rating of not less than 660 watts if of the admedium type, or not less than 750 watts if of any other type.

       A. high voltage
       B. high amperage
       C. heavy-duty type
       D. rough use

Timed Exam 3-30   All wet niche swimming pool luminaries shall be removable from the water for relamping or normal maintenance. Luminaries shall be installed in such a manner that personnel can reach the luminaire for relamping, maintenance, or inspection while on the deck or equivalently _____.

       A. wet location
       B. dry location
       C. accessible location
       D. none of the above

Timed Exam 3-31   Underground service conductors shall be insulated for _____.

       A. workers safety
       B. expose to the elements
       C. the applied voltage
       D. identification purposes

Timed Exam 3-32   Locations of lamps for outdoor lighting shall be below all energized conductors, transformers, or other electric utilization equipment, unless either of the following apply

       A. either C or D
       B. Equipment is controlled by a disconnecting means that can be locked in the closed position.
       C. Equipment is controlled by a disconnecting means that is lockable in accordance with 110.25.
       D. Clearances or other safeguards are provided for relamping operations.

Timed Exam 3-33   Overhead spans of open conductors and open multiconductor cables of not over 1000 volts, nominal, shall have a clearance of not less than _____ ft above finished grade, sidewalks, or from any platform or projection from which they might be reached where the voltage does not exceed 150 volts to ground and accessible to pedestrians only

       A. 10
       B. 12
       C. 15
       D. 18

Timed Exam 3-34   Where used outside, aluminum or copper-clad aluminum grounding electrode conductors shall not be terminated within _____ in. of the earth.

       A. 12
       B. 18
       C. 24
       D. 36

Timed Exam 3-35   What is Type HFF wiring used for.

       A. Heat, Fire and Flame resistant
       B. underground
       C. Fixture wiring
       D. hydrogen atmosphere applications

Timed Exam 3-36   With respect to Solar Photovoltaic (PV) Systems equipment, overcurrent device ratings shall be not less than _____ of the maximum currents calculated in 690.8(A)., Exceptioned ignored.

       A. 100 percent
       B. 125 percent
       C. 150 percent
       D. 200 percent

Timed Exam 3-37 To prevent corrosion, raceways, cable trays, cablebus, auxiliary gutters, cable armor, boxes, cable sheathing, cabinets, elbows, couplings, fit-tings, supports, and support hardware shall be of materials suitable for the _____ in which they are to be installed.

        A. location
        B. position
        C. environment
        D. humidity

Timed Exam 3-38 Means shall be provided for disconnecting all _____ conductors that supply or pass through the building or structure.

        A. grounded
        B. ungrounded
        C. grounding
        D. bonding

Timed Exam 3-39 There shall not be more than the equivalent of four quarter bends _____ degrees total between pull points, for example, conduit bodies and boxes.

        A. 180
        B. 270
        C. 360
        D. 450

Timed Exam 3-40 What is the maximum fill allowance for a 4 inch square box 2.125" deep, for No. 12 copper conductors.

        A. 10
        B. 12
        C. 13
        D. 15

Timed Exam 3-41   As pretaining to class 1, class 2, and class 3 remote-control, signaling, and power-limited circuits, separation of ground-fault protection time-current characteristics shall conform to the manufacturer's recommendations and shall consider all required tolerances and disconnect operating time to achieve _____ selectivity.

        A. 50 percent
        B. 75 percent
        C. 100 percent
        D. none of these

Timed Exam 3-42   Electrical Nonmetallic Tubing, ENT may be used _____. Exceptions ignored.

        A. Where the voltage is over 600 volts
        B. Where subject to physical damage
        C. In a two story building
        D. For direct earth burial

Timed Exam 3-43   The overhead conductors between the utility electric supply system and the service point.

        A. Service Entrance
        B. Service Drop
        C. Service Lateral
        D. Power Entrance

Timed Exam 3-44   Which of the following areas of construction does the NEC (National Electrical Code) not cover?

        A. private premises
        B. parking lots
        "C. automotive vehicles other than mobile homes and recreational vehicles"
        D. Installations used by the electric utility

Timed Exam 3-45   In dwelling units, where two or more branch circuits supply devices or equipment on the same yoke or mounting strap, a means to _____ the ungrounded conductors supplying those devices shall be provided at the point at which the branch circuits originate.

        A. connect
        B. simultaneously disconnect
        C. disconnect
        D. energize

Timed Exam 3-46   The greatest root-mean-square (rms) (effective) difference of potential between any two conductors of the circuit concerned.

        A. Amperage
        B. Wattage
        C. Voltage
        D. Power Factor

Timed Exam 3-47   Aluminum, copper-clad aluminum, or copper conductors of size 1/0 AWG and larger, comprising each phase, polarity, neutral, or grounded circuit conductor, shall be permitted to be connected in _____ (electrically joined at both ends). Exceptions Ignored.

        A. parallel
        B. series
        C. delta
        D. wye

Timed Exam 3-48   Up to three sets of _____ or two sets of 4-wire or 5-wire feeders shall be permitted to utilize a common neutral.

        A. 2-wire feeders
        B. insulated conductors
        C. 3-wire feeders
        D. 100 amp feeders

Timed Exam 3-49   Any nonconductive paint, enamel, or similar coating shall be _____ at threads, contact points, and contact surfaces or be connected by means of fittings designed so as to make such removal unnecessary.

      A. removed
      B. tested
      C. continuity checked
      D. acid etched

Timed Exam 3-50   A warehouse storage area uses mercury vapor lighting. What is the minimum lighting load in the warehouse area if it has 40,500 sq. ft. for storage.

      A. 10125 VA
      B. 7875 VA
      C. 6625 VA
      D. 5165 VA

Timed Exam 3-51   Equipment intended to interrupt current at fault levels shall have an interrupting ratingat nominal circuit _____ sufficient for the current that is available at the line terminals of the equipment.

      A. voltage
      B. power
      C. resistance
      D. size

Timed Exam 3-52   All electric pool water heaters shall have the heating elements subdivided into loads not exceeding _____ amperes and protected at not over _____ amperes.

      A. 20, 40
      B. 40, 80
      C. 48, 60
      D. 50, 80

Timed Exam 3-53   In dwelling units, where two or more single-phase ranges are supplied by a 3-phase, 4-wire feeder or service, the total load shall be calculated on the basis of _____ the maximum number connected between any two phases.

> A. one and one half
> B. twice
> C. three times
> D. four times

Timed Exam 3-54   Each conductor that originates outside the box and terminates or is spliced within the box shall be counted once, and each conductor that passes through the box without splice or termination _____.

> A. does not have to be counted
> B. shall be counted twice
> C. shall be counted once
> D. shall be counted once for all such conductors as a group

Timed Exam 3-55   The disconnecting means shall be installed _____ of the building or structure served or where the conductors pass through the building or structure and, the disconnecting means shall be at a readily accessible location nearest the point of entrance of the conductors. Exceptions ignored.

> A. inside
> B. outside
> C. inside and outside
> D. either inside or outside

Timed Exam 3-56   All non-current-carrying metal parts of electric equipment and all metal raceways and cable sheaths shall be solidly grounded and bonded to all metal pipes and rails at the portal and at intervals not exceeding _____ ft throughout a tunnel.

> A. 500
> B. 1000
> C. 1500
> D. 2000

Timed Exam 3-57   The grounding electrode conductor shall be of _____. or the items as permitted in 250.68(C).

       A. copper
       B. aluminum
       C. copper-clad aluminum
       D. any of the above

Timed Exam 3-58   Which of the following is not considered a special purpose branch circuit as outlined in the code.

       A. Circuits and equipment operating at less than 50 volts
       B. Cranes and hoists
       C. High pressure sodium parking lot lighting
       D. Fixed electric space-heating equipment

Timed Exam 3-59   Handles or levers of circuit breakers, and similar parts that may move suddenly in such a way that persons in the vicinity are likely to be injured by being struck by them, shall _____.

       A. not be used
       B. be made non-accessible
       C. locked in the open position
       D. be guarded or isolated

Timed Exam 3-60   What is the minimum bending space at a terminal required for a 4/0 copper conductor with 3 wires per terminal.

       A. 5.5 inches
       B. 8.5 inches
       C. 9 inches
       D. 10 inches

Timed Exam 3-61   Conductors in raceways shall be _____ between outlets, boxes, devices, and so forth. Special exceptions ignored.

       A. pulled
       B. installed
       C. continuous
       D. in sections

Timed Exam 3-62   Switches shall not be installed within tub or shower spaces _____.

      A. unless they are grounded, bonded and GFI protected
      B. unless they have a watertight cover
      C. unless installed as part of a listed tub or shower assembly.
      D. under any circumstances

Timed Exam 3-63   A device that establishes a connection between two or more conductors or between one or more conductors and a terminal by means of mechanical pressure and without the use of solder.

      A. Solder Connector
      B. Type P Connector
      C. Pressure Connector
      D. Wire Splice

Timed Exam 3-64   A complete lighting unit consisting of a light source such as a lamp or lamps, together with the parts designed to position the light source and connect it to the power supply. It may also include parts to protect the light source or the ballast or to distribute the light.

      A. Light Fixture
      B. Incandescent Lamp
      C. Street Lamp
      D. Luminaire

Timed Exam 3-65   What is the allowable ampacity for a No. 14 fixture wire.

      A. 15
      B. 17
      C. 20
      D. 25

Timed Exam 3-66 A multi-family dwelling has 40 dwelling units. Meters are in two banks of 20 each with individual feeders to each dwelling unit. One-half of the dwelling units are equipped with electric ranges not exceeding 12 kW each. Assume range kW rating equivalent to kVA rating in accordance with 220.55. Other half of ranges are gas ranges. Area of each dwelling unit is 840 sq. ft. Laundry facilities on premises are available to all tenants. Add no circuit to individual dwelling unit. Could you use 2 15 A. circuits to handle the general lighting load?

       A. Yes
       B. No

Timed Exam 3-67 Type MV cable shall be permitted for use on power systems rated up to and including _____ volts nominal.

       A. 600
       B. 2000
       C. 10,000
       D. 35,000

Timed Exam 3-68 For residential branch circuits rated 120 volts or less with GFIC protection and maximum overcurrent protection of 20 amperes, What is the minimum burial depth In inches in a trench below 2-in. thick concrete or equivalent.

       A. 6
       B. 12
       C. 18
       D. 24

Timed Exam 3-69 A device or group of devices that serves to govern, in some predetermined manner, the electric power delivered to the apparatus to which it is connected.

       A. Remote Relay
       B. Controller
       C. Circuit Breaker
       D. Governor

Timed Exam 3-70 Conductors of branch circuits supplying more than one receptacle for _____ loads shall have an ampacity of not less than the rating of the branch circuit.

    A. permanent
    B. most
    C. cord-and-plug-connected portable
    D. washing machine

Timed Exam 3-71 An assembly of a fuse support with either a fuse-holder, fuse carrier, or disconnecting blade. The fuseholder or fuse carrier may include a conducting element (fuse link) or may act as the disconnecting blade by the inclusion of a nonfusible member.

    A. Disconnect
    B. Cutout
    C. Breaker
    D. Fused Switch

Timed Exam 3-72 The branch-circuit rating for an appliance that is a continuous load, other than a motor-operated appliance, shall not be less than _____ percent of the marked rating, or not less than 100 percent of the marked rating if the branch-circuit device and its assembly are listed for continuous loading at 100 percent of its rating.

    A. 100
    B. 125
    C. 150
    D. 175

Timed Exam 3-73 A single receptacle installed on an individual branch circuit shall have an ampere rating of _____ that of the branch circuit.

    A. not less than 100% of
    B. not less than 50% of
    C. not less than 80% of
    D. not less than 125% of

Timed Exam 3-74 _____ shall not be used where conduits or connectors requiring the use of locknuts or bushings are to be connected to the side of the box.

    A. Handy boxes
    B. Octagon boxes
    C. Square boxes
    D. Round boxes

Timed Exam 3-75   General-use dimmer switches can be used to control permanently installed unless listed for the control of other loads and installed accordingly.

    A. paddle fans
    B. fluorescent lights
    C. lighting receptacle outlets
    D. incandescent luminaries

Timed Exam 3-76   Disconnecting means shall be _____. The provisions for locking shall remain in place with or without the lock installed.

    A. locked from public tampering
    B. lockable in accordance with 110.25.
    C. locked in the closed position
    D. A and C

Timed Exam 3-77   The grounding electrode conductor shall be _____. or the items as permitted in 250.68(C).

    A. copper, aluminum, or copper-clad aluminum
    B. solid or stranded, and insulated
    C. stranded, insulated, covered, or bare
    D. solid or stranded, and bare

Timed Exam 3-78   Circuit breakers shall be marked with their ampere rating in a manner that will be durable and visible after installation, the marking shall be permitted to be made visible by _____.

       A. tripping the breaker
       B. removal of a trim or cover
       C. removing the breaker
       D. markings on the cover

Timed Exam 3-79   For Fire alarm circuits, circuit integrity (CI) cables shall be supported at a distance not exceeding _____. Where located within _____ of the floor, as covered in 760.53(A)(1) and 760.130(1), as applicable, the cable shall be fastened in an approved manner at intervals of not more than _____. Cable supports and fasteners shall be steel

       A. 48 in. , 7 ft , 18 in.
       B. 24 in. , 2 ft , 18 in.
       C. 24 in. , 7 ft , 18 in.
       D. 24 in. , 7 ft , 36 in.

Timed Exam 3-80   Where a building or structure has any combination of feeders, branch circuits, or services _____, a permanent plaque or directory shall be installed at each feeder and branch-circuit disconnect location that denotes all other services, feeders, or branch circuits supplying that building or structure or passing through that building or structure and the area served by each.

       A. passing through or supplying it
       B. passing through it
       C. supplying it
       D. all of the above makes this statement true

Timed Exam 4   3 1/2 Hours to Complete

Timed Exam 4-1 _____ meeting the requirements of this article shall be used around impaired connections, such as reducing washers or oversized, concentric, or eccentric knockouts. Standard locknuts or bushings shall not be the only means for the bonding required by this section but shall be permitted to be installed to make a mechanical connection of the raceway(s).

    A. Locknuts
    B. Copper bushings
    C. Compression fittings
    D. Bonding jumpers

Timed Exam 4-2   With respect to Solar Photovoltaic (PV) Systems equipment, where energy storage device input and output terminals are more than _____ from connected equipment, or where the circuits from these terminals pass through a wall or partition, the installation shall comply with the following: (1) A disconnecting means and overcurrent protection shall be provided at the energy storage device end of the circuit. Fused disconnecting means or circuit breakers shall be permitted to be used. (2) Where fused disconnecting means are used, the line terminals of the disconnecting means shall be con- nected toward the energy storage device terminals. (3) Overcurrent devices or disconnecting means shall not be installed in energy storage device enclosures where explosive atmospheres can exist. (4) A second disconnecting means located at the connected equipment shall be installed where the disconnecting means required by 690.71(H)(1) is not within sight of the connected equipment. (5) Where the energy storage device disconnecting means is not within sight of the PV system ac and dc discon- necting means, placards or directories shall be installed at the locations of all disconnecting means indicating the location of all disconnecting means.

    A. 3 ft
    B. 5 ft
    C. 10 ft
    D. 15 ft

Timed Exam 4-3   The length of the cord for a kitchen waste disposer shall not be less than _____ in. and not more than _____ in.

    A. 24, 42
    B. 30, 48
    C. 36, 72
    D. 18, 36

Timed Exam 4-4   For feeders over 600 Volts, the ampacity of feeder conductors shall not be less than _____.

    A. 80 Amps
    B. 125% of the sum of the nameplate ratings of the transformers supplied when only transformers are supplied
    C. the sum of the nameplate ratings of the transformers supplied when only transformers are supplied
    D. 150% of the sum of the nameplate ratings of the transformers supplied when only transformers are supplied

Timed Exam 4-5   The use of EMT shall be permitted for _____ work.

    A. underwater
    B. exposed
    C. concealed
    D. both B and C

Timed Exam 4-6   Two-wire dc circuits and ac circuits of two or more _____ shall be permitted to be tapped from the ungrounded conductors of circuits having a grounded neutral conductor. Switching devices in each tapped circuit shall have a pole in each ungrounded conductor.

    A. grounded conductors
    B. ungrounded conductors
    C. ungrounding conductors
    D. grounding conductors

Timed Exam 4-7   Conductors, other than _____, shall be protected against overcurrent in accordance with their ampacities.

    A. service lateral conductors
    B. heating overload conductors
    C. flexible cords, flexible cables, and fixture wires
    D. variable gauge conductors

Timed Exam 4-8   Bends in Types NM, NMC, and NMS cable shall be so made that the cable will not be damaged. The radius of the curve of the inner edge of any bend during or after installation shall not be less than _____ times the diameter of the cable.

    A. ten
    B. twenty
    C. five
    D. thirty six

Timed Exam 4-9   Parts of electric equipment that in ordinary operation produce arcs, sparks, flames, or molten metal shall be enclosed or separated and isolated from all _____.

    A. other electrical components
    B. possible contact with people
    C. combustible material
    D. metal parts

Timed Exam 4-10   How much free space in cubic inches is required in a octagon box that has 6 - No. 12 wires entering and terminating in it. There is also 3 - grounding conductors terminating in it. There is also a fixture stud and three entry clamps inside of the box.

    A. 18.5
    B. 20
    C. 20.25
    D. 21.75

Timed Exam 4-11   Where a feeder overcurrent device is not readily accessible, branch-circuit overcurrent devices shall be installed on the load side, shall be mounted in a readily accessible location, and shall be of _____ ampere rating than the feeder overcurrent device.

        A. a lower
        B. a higher
        C. the same
        D. 150 percent of the

Timed Exam 4-12   In locations where electric equipment is likely to be exposed to physical damage, enclosures or _____ shall be so arranged and of such strength as to prevent such damage.

        A. insulation
        B. guards
        C. conductors
        D. equipment

Timed Exam 4-13   The means of coupling to the electric vehicle shall be either conductive or inductive. At- tachment plugs, electric vehicle connectors, and electric vehicle inlets shall be listed or labeled for the purpose. The overall usable length shall not exceed _____ unless equipped with a cable management system that is part of the listed electric vehicle supply equipment.

        A. 8 ft
        B. 15 ft
        C. 25 ft
        D. 50 ft

Timed Exam 4-14   All _____ of the same circuit and, where used, the grounded conductor and all equipment grounding conductors and bonding conductors shall be contained within the same raceway, auxiliary gutter, cable tray, cablebus assembly, trench, cable, or cord. Exceptions excluded.

        A. control conductors
        B. relay conductors
        C. conductors
        D. communication

Timed Exam 4-15 The disconnecting means for each supply shall consist of not more than _____ switches or _____ circuit breakers mounted in a single enclosure, in a group of separate enclosures, or in or on a switchboard or switchgear.

        A. six, six
        B. two, two
        C. three, three
        D. eight, eight

Timed Exam 4-16 Which of the following shall not be used as grounding electrodes:

        A. Metal underground gas piping system
        B. Aluminum electrodes
        C. Concrete rebar
        D. A and B

Timed Exam 4-17 Unless required elsewhere in this Code, equipment grounding conductors shall be permitted to be bare, covered, or _____. Exceptions ignored.

        A. painted green
        B. energized
        C. insulated
        D. fused

Timed Exam 4-18 In a dwelling unit the back of a non-corner sink is located 14" from a wall, does the code require an outlet to be installed in that area if the distance to the nearest countertop outlet is 3 feet.

        A. Yes
        B. No

Timed Exam 4-19   When calculating conductor fill in a box, each loop or coil of, unbroken conductor not less than twice the minimum length required for free conductors shall be counted _____.

        A. as zero
        B. once
        C. as three
        D. twice

Timed Exam 4-20   A 125-volt, single-phase, 15- or 20-ampere-rated receptacle outlet shall be installed at an accessible location for the servicing of heating, air-conditioning, and refrigeration equipment, and the receptacle shall be located on the same level and within _____ of the heating, air-conditioning, and refrigeration equipment.

        A. 25 ft
        B. 15 ft
        C. 10 ft
        D. 6 ft

Timed Exam 4-21   Multiconductor cables used for overhead service conductors shall be attached to buildings or other structures by fittings identified for use with service conductors. Open conductors shall be attached to fittings identified for use with service conductors or to noncombustible, _____ insulators securely attached to the building or other structure.

        A. glass
        B. metallic
        C. plastic
        D. nonabsorbent

Timed Exam 4-22   Openings through which conductors enter a box shall be _____.

        A. sealed
        B. clamped shut
        C. open for ventilation
        D. closed in an approved manner

Timed Exam 4-23 An apparatus designed to control and organize unused lengths of output cable to the electric vehicle.

    A. Energy conservation control system (Electric Vehicle Supply Equipment)
    B. Energy Management control board system (Electric Vehicle Supply Equipment)
    C. Green vehicle CO2 footprint optimization system (Electric Vehicle Supply Equipment)
    D. Cable Management System (Electric Vehicle Supply Equipment)

Timed Exam 4-24 Liquidtight Flexible Metal Conduit (LFMC) is permitted to be used _____.

    A. Where total bends exceeds 360 degrees.
    B. For direct burial where listed and marked for the purpose
    C. Where subject to physical damage
    D. Where any combination of ambient and conductor temperature produces an operating temperature in excess of that for which the material is approved

Timed Exam 4-25 Overhead spans of open conductors and open multiconductor cables shall have a vertical clearance of not less than ____ ft above the roof surface. The vertical clearance above the roof level shall be maintained for a distance not less than ft in all directions from the edge of the roof. Exceptions ignored.

    A. 8, 3
    B. 12, 6
    C. 12, 4
    D. 8, 10

Timed Exam 4-26 In dwelling unit garages. In each attached garage and in each detached _____. The branch circuit supplying this receptacle(s) shall not supply outlets outside of the garage. At least one receptacle outlet shall be installed for each car space.

    A. detached garage or accessory building with electric power
    B. walkway to the garage
    C. 6' space along the floorline
    D. 12' or wider carport

Timed Exam 4-27   All 120-volt, single phase, 15- and 20-ampere branch circuits supplying outlets or devices installed in dwelling unit kitchens, family rooms, dining rooms, living rooms, parlors, libraries, dens, bedrooms, sunrooms, recreation rooms, closets, hallways, or similar rooms or areas shall be protected by a listed _____ installed to provide protection of the branch circuit.

    A. ground fault interupter
    B. fuse
    C. arc-fault circuit interrupter, combination-type,
    D. bonded jumper

Timed Exam 4-28   All wiring from the controllers to fire pump motors shall be in rigid metal conduit, intermediate metal conduit, electrical metallic tubing, liquidtight flexible metal conduit, or liquidtight flexible nonmetallic conduit Type LFNC-B, listed Type MC cable with an impervious covering, or Type MI cable. Electrical connections at motor terminal boxes shall be made with a listed means of connection such as _____.

    A. twist-on type
    B. insulation-piercing type
    C. soldered wire connectors
    D. none of the above

Timed Exam 4-29   Metal or nonmetallic raceways, cable armors, and cable sheaths shall be _____ between cabinets, boxes, fit-tings, or other enclosures or outlets.

    A. continuous
    B. anchored
    C. supported
    D. horizontal

Timed Exam 4-30   What is the maximum allowable ampacity for No. 12 copper conductors type THHN, 167 F rated insulation, installed in an ambient temperature of 30 C. with 27 current carrying conductors in the same raceway.

       A. 25
       B. 15
       C. 11.25
       D. 18.75

Timed Exam 4-31   For permanently connected appliances rated at not over 300 volt-amperes or 1/8hp, the branch-circuit overcurrent device shall _____.

       A. be within sight of the appliance
       B. not be permitted to serve as the disconnecting means
       C. be permitted to serve as the disconnecting means
       D. none of the above

Timed Exam 4-32   What is the ampacity rating of No. 1 (AWG) copper-clad aluminum conductors at 90 C. in multiconductor cables with not more than three insulated conductors, rated 0 through 2000 Volts, in free air based on ambient air temperature of 40°C for type MC cables

       A. 84
       B. 108
       C. 120
       D. 126

Timed Exam 4-33   Where a building or other structure that is served by a branch circuit or feeder on the load side of a service disconnecting means shall be supplied by only _____. Ignoring all special conditions.

       A. only one feeder or branch circuit
       B. a separate service
       C. multiple branch circuit feeders
       D. new service entrance cabling

Timed Exam 4-34 Underground cable and conductors installed _____ shall be in a raceway. Exceptions ignored.

        A. in a basement
        B. under a building
        C. in an attic
        D. in a floor

Timed Exam 4-35 _____ shall be provided to give safe access to the working space around electric equipment installed on platforms, balconies, or mezzanine floors or in attic or roof rooms or spaces.

        A. Elevators
        B. Permanent ladders or stairways
        C. Temporary ladders
        D. Collapsible stairways

Timed Exam 4-36 Conduits or raceways through which moisture may contact live parts shall be _____.

        A. RMC
        B. RMC or RNC
        C. EMT, RMC, or RNC
        D. sealed or plugged at either or both ends

Timed Exam 4-37 A dwelling has a floor area of 1500 sq. ft, with unfinished cellar not adaptable for future use, unfinished attic, and open porches. Appliances are a 12-kW range and a 5.5-kW, 240-V dryer. Assume range and dryer kW ratings equivalent to kVA ratings in accordance with 220.54 and 220.55. What is the general lighting load in VA units.

        A. 2700 VA
        B. 3800 VA
        C. 4500 VA
        D. 6000 VA

Timed Exam 4-38   Where a conduit enters a box, fitting, or other enclosure, _____ shall be provided to protect the wire from abrasion unless the box, fitting, or enclosure design provides equivalent protection.

       A. a bushing or adapter
       B. a insulated collar
       C. pipe nipple
       D. end clamp

Timed Exam 4-39   With respect to Solar Photovoltaic (PV) Systems equipment, required ground fault protection devices or systems shall:

       A. Be capable of detecting a ground fault in the PV array dc current-carrying conductors and components, in- cluding any intentionally grounded conductors,
       B. Interrupt the flow of fault current
       C. Provide an indication of the fault
       D. Be listed for providing PV ground-fault protection
       E. All of the above

Timed Exam 4-40   Where conductors are run in parallel in multiple raceways or cables, the equipment grounding conductors, where used, shall be run _____.

       A. in one of the conductor raceways or cables
       B. in parallel in each raceway or cable
       C. a separate raceway or cable
       D. any of the above

Timed Exam 4-41   Cellular Concrete Floor Raceways shall not be used In commercial garages, other than for supplying _____ outlets or extensions to the area below the floor but not above.

       A. floor
       B. wall
       C. ceiling
       D. roof

Timed Exam 4-42 Overcurrent devices shall be readily accessible and shall be installed so that the center of the grip of the operating handle of the switch or circuit breaker, when in its highest position, is not more than 6 ft ___ in. above the floor or working platform. Exceptions ignored.

    A. 6
    B. 7
    C. 8
    D. 10

Timed Exam 4-43 A concrete-encased electrode shall consist of at least 6.0 m (20 ft) of either (1) or (2): (1) One or more bare or zinc galvanized or other electrically conductive coated steel reinforcing bars or rods of not less than 13 mm (12 in.) in diameter, installed in one continuous _____ foot length, or if in multiple pieces connected together by the usual steel tie wires, exothermic welding, welding, or other effective means to create a 6.0 m (20 ft) or greater length; or (2) Bare copper conductor not smaller than 4 AWG Metallic components shall be encased by at least _____ inches of concrete and shall be located horizontally within that portion of a concrete foundation or footing that is in direct contact with the earth or within vertical foundations or structural components or members that are in direct contact with the earth. If multiple concrete-encased electrodes are present at a building or structure, it shall be permissible to bond only one into the grounding electrode system.

    A. 2,16
    B. 3, 20
    C. 4, 20
    D. 20, 2

Timed Exam 4-44 An enclosure that is designed for either surface mounting or flush mounting and is provided with a frame, mat, or trim in which a swinging door or doors are or can be hung.

    A. Panel Board
    B. Controller
    C. Cabinet
    D. Cutout Box

Timed Exam 4-45   For circuits supplying lighting units that have ballasts, transformers, or autotransformers, the calculated load shall be based on _____

       A. total watts of the lamps
       B. efficiency ratio of the luminaire
       C. the total ampere ratings of such units and not on the total watts of the lamps
       D. number of lamps per unit x the wattage of each lamp

Timed Exam 4-46   Ground Fault

       A. Connected (connecting) to ground or to a conductive body that extends the ground connection.
       B. Connected to ground without inserting any resistor or impedance device.
       C. An unintentional, electrically conducting connection between an ungrounded conductor of an electrical circuit and the normally non-current-carrying conductors, metallic enclosures, metallic raceways, metallic equipment, or earth.
       D. The circuit conductors between the final overcurrent device protecting the circuit and the outlet(s).

Timed Exam 4-47   Conductors shall be permitted to be terminated based on the ____°C temperature rating and ampacity as given in Table 310.60(C)(67) through Table 310.60(C)(86), unless otherwise identified.

       A. 60
       B. 75
       C. 90
       D. 100

Timed Exam 4-48   The dwelling has a floor area of 1500 sq. ft. exclusive of an unfinished cellar not adaptable for future use, unfinished attic, and open porches. It has two 20-A small appliance circuits, one 20-A laundry circuit, two 4-kW wall-mounted ovens, one 5.1-kW counter-mounted cooking unit, a 4.5-kW water heater, a 1.2-kW dishwasher, a 5-kW combination clothes washer and dryer, six 7-A, 230-V room air-conditioning units, and a 1.5-kW, 230 V permanently installed bathroom space heater. Assume wall-mounted ovens, counter-mounted cooking unit, water heater, dishwasher, and combination clothes washer and dryer kW ratings equivalent to kVA. What is the total calculated load for the service in VA and the total required minimum service rating, assuming that the two 4-kVA wall-mounted ovens are supplied by one branch circuit, the 5.1-kVA counter-mounted cooking unit by a separate circuit.

   A. Calculated Service load 29,200 VA - Minimum Service Rating 122A

   B. Calculated Service load 15,200 VA - Minimum Service Rating 100A

   C. Calculated Service load 39,200 VA - Minimum Service Rating 150A

   D. Calculated Service load 5,200 VA - Minimum Service Rating 60A

Timed Exam 4-49   _____ connectors shall not be concealed.

   A. rigid
   B. XLT
   C. angle
   D. 45 degree offset

Timed Exam 4-50   At least _____ entrance of sufficient area shall be provided to give access to and egress from working space about electrical equipment.

   A. 1
   B. 2
   C. 3
   D. 4

Timed Exam 4-51 A compartment or chamber to which one or more air ducts are connected and that forms part of the air distribution system.

    A. Heat Exchange
    B. Plenum
    C. Duct
    D. Air Vent

Timed Exam 4-52 Wire connectors or splicing means installed on conductors for direct burial shall _____.

    A. be listed for such use
    B. not be used
    C. encased in concrete
    D. made waterproof

Timed Exam 4-53 One method of marking an insulated grounded conductor larger than 6 AWG shall be _____.

    A. by using a bare copper conductor
    B. by noting which terminal to which it is connected
    C. At the time of installation, by a distinctive white or gray marking at its terminations. This marking shall encircle the conductor or insulation.
    D. by a distinctive green marking encircling the conductor or insulation at its terminations installed at the time of installation.

Timed Exam 4-54 Splices or taps shall be permitted within gutters where they are accessible by means of removable covers or doors. The conductors, including splices and taps, shall not fill the gutter to more than _____ percent of its area.

    A. 25
    B. 50
    C. 75
    D. 90

Timed Exam 4-55   For circuits supplying loads consisting of motor-operated utilization equipment that is fastened in place and has a motor larger than ____ hp in combination with other loads, the total calculated load shall be based on _____ percent of the largest motor load plus the sum of the other loads.

      A. 1/8, 125
      B. 1/4, 110
      C. 1/2, 150
      D. 3/4, 80

Timed Exam 4-56   Cabinets and cutout boxes shall have approved space to accommodate all conductors installed in them without _____.

      A. bending
      B. crowding
      C. kinking
      D. trimming

Timed Exam 4-57   Rod and pipe grounding electrodes shall not be less than _____ ft in length and shall consist of the code specified materials.

      A. 4
      B. 6
      C. 8
      D. 10

Timed Exam 4-58   Lighting busway and trolley busway shall not be installed less than _____ ft above the floor or working platform unless provided with a cover identified for the purpose.

      A. 8
      B. 10
      C. 12
      D. 14

Timed Exam 4-59   Ground-fault circuit-interrupter protection for personnel shall be provided for outlets that supply boat hoists installed in dwelling unit locations and supplied by _____.

    A. all 125-volt, 15- and 20-ampere or greater branch circuits
    B. all 20-ampere branch circuits
    C. outlets not exceeding 240 volts
    D. three wire branch circuits 125-volt, 15- and 20-ampere

Timed Exam 4-60   An uninsulated equipment grounding conductor shall be permitted, but, if individually covered, the covering shall have a continuous outer finish that is _____. Exceptions ignored.

    A. green
    B. green with yellow strips
    C. yellow with green strips
    D. A or B

Timed Exam 4-61   Type UF cable may be used as follows:

    A. In storage battery rooms
    B. As service-entrance cable
    C. Between a dwelling and separate garage unit
    D. Where exposed to direct rays of the sun, unless identified as sunlight resistant

Timed Exam 4-62   In dwelling units the lighting demand factor of the first 3000 VA or less is _____ Percent.

    A. 50
    B. 80
    C. 100
    D. 125

Timed Exam 4-63   What is the maximum allowable distance between trade size 1 RMC supports in feet.

       A. 10
       B. 12
       C. 14
       D. 16

Timed Exam 4-64   Overhead conductors for festoon lighting shall not be smaller than 12 AWG unless the conductors are supported by messenger wires. In all spans exceeding _____ ft, the conductors shall be supported by messenger wire.

       A. 40
       B. 50
       C. 60
       D. 100

Timed Exam 4-65   For a one-family dwelling, the feeder disconnecting means shall have a rating of not less than _____ amperes, 3-wire.

       A. 60
       B. 90
       C. 100
       D. 125

Timed Exam 4-66   A conductor, other than a service conductor, that has overcurrent protection ahead of its point of supply that exceeds the value permitted for similar conductors that are protected as described elsewhere in 240.4.

       A. branch circuit conductor
       B. ungrounded conductor
       C. tap conductor
       D. overload conductor

Timed Exam 4-67   An electric power production system that is operating in parallel with and capable of delivering energy to an electric primary source supply system.

        A. Motor Control Center
        B. Multioutlet Assembly
        C. Interactive System
        D. Supplementary Overcurrent Protective Device

Timed Exam 4-68   Normally non-current-carrying conductive materials enclosing electrical conductors or equipment, or forming part of such equipment, shall be connected together and to the electrical supply source in a manner that establishes an effective ground-fault current path.

        A. bonding of electrical equipment.
        B. grounded conductor
        C. ungrounded conductor
        D. ground fault

Timed Exam 4-69   Two or more grounding electrodes that are effectively bonded together shall be considered _____.

        A. a multiple grounding electrode system
        B. a continuous grounding electrode system.
        C. a single grounding electrode system
        D. a redundant grounding electrode system.

Timed Exam 4-70   For cord-and-plug-connected equipment in other than storable pools, the flexible cords shall not exceed _____ ft in length.

        A. 3
        B. 6
        C. 10
        D. 15

Timed Exam 4-71 A manually operated device used in conjunction with a transfer switch to provide a means of directly connecting load conductors to a power source and of disconnecting the transfer switch.

       A. Bypass Isolation Switch
       B. Transfer Switch
       C. Double Pole Switch
       D. 4 Way Switch

Timed Exam 4-72 The point of attachment of the overhead service conductors to a building or other structure shall provide the minimum clearances an in no case shall this point of attachment be less than _____ ft above finished grade.

       A. 8
       B. 10
       C. 12
       D. 14

Timed Exam 4-73 Service cables, where subject to physical damage, shall be protected by any of the following except.

       A. Rigid metal conduit
       B. Schedule 80 rigid PVC conduit
       C. Electrical metallic tubing
       D. Intermediate metal conduit
       E. all are acceptable

Timed Exam 4-74 What is the approximate diameter in inches of No. 2 RHW wire.

       A. .412
       B. .375
       C. .275
       D. .125

Timed Exam 4-75 Electrodes of bare or conductively coated iron or steel plates must be at least _____ in. thick and at least _____ sq. feet in contact with exposed earth to be used as a grounding electrode.

        A. 0.06, 3
        B. 0.06, 2
        C. .25, 2
        D. 0.06, 4

Timed Exam 4-76 Nonconductive coatings (such as paint, lacquer, and enamel) on equipment to be grounded shall be removed from threads and other contact surfaces to ensure good electrical _____ or be connected by means of fittings designed so as to make such removal unnecessary.
        A. conduction
        B. voltage
        C. current
        D. continuity

Timed Exam 4-77 Receptacle outlets in or on floors shall not be counted as part of the required number of receptacle outlets unless located within _____ of the wall.

        A. 1 ft.
        B. 1.5 ft.
        C. 2 ft.
        D. 8 in.

Timed Exam 4-78 The equipment grounding conductor run with or enclosing the circuit conductors may be _____ conductor.

        A. copper, aluminum, or copper-clad aluminum
        B. copper
        C. aluminum
        D. copper-clad aluminum

Timed Exam 4-79 The NEC (National Electrical Code) is not intended as a design specification or an instruction manual for _____ .

        A. electrical engineers
        B. untrained persons
        C. power linemen
        D. none of the above

Timed Exam 4-80   In dwelling units, no small-appliance branch circuit shall _____.

       A. serve more than two kitchens
       B. serve more than one dining room
       C. serve a more than one kitchen
       D. serve a dining room and a kitchen

Timed Exam 5   3 1/2 Hours to Complete

Timed Exam 5-1   Type MV cable shall not be used _____, unless identified for the use.

    A. in wet or dry locations
    B. in raceways
    C. where exposed to direct sunlight
    D. in messenger-supported wiring

Timed Exam 5-2   Flexible Metallic Tubing (FMT) can be used in which of the following places or conditions.

    A. In hoistways
    B. Under ground for direct earth burial, or embedded in poured concrete or aggregate
    C. For system voltages of 1000 volts maximum
    D. In lengths over 100 ft

Timed Exam 5-3   Metal raceways, cable trays, cable armor, cable sheath, enclosures, frames, fittings, and other metal noncurrent-carrying parts that are to serve as equipment grounding conductors, with or without the use of supplementary equipment grounding conductors, shall be effectively bonded where necessary to ensure _____ and the capacity to conduct safely any fault current likely to be imposed on them.

    A. ground stability
    B. electrical continuity
    C. electrical insulation
    D. electrical inductance

Timed Exam 5-4   Does a pipe organ have any special grounding requirements?

    A. Yes
    B. No

Timed Exam 5-5   A grounding rod electrode shall be permitted to be buried in a trench that is at least _____ in. deep, and the upper end of the electrode shall be flush with or below ground level unless the aboveground end and the grounding electrode conductor attachment are protected against physical damage as specified in 250.10.

        A. 60
        B. 48
        C. 36
        D. 30

Timed Exam 5-6   The normally non-current-carrying metal parts of all service enclosures shall be effectively _____.

        A. bonded together
        B. insulated
        C. protected
        D. locked

Timed Exam 5-7   The minimum branch-circuit conductor size shall have an allowable ampacity not less than the _____ load to be served after the application of any adjustment or correction factors

        A. maximum
        B. minimum
        C. constant
        D. intermittent

Timed Exam 5-8   Where practicable, rod, pipe, and plate electrodes shall be embedded below _____. Rod, pipe, and plate electrodes shall be free from nonconductive coatings such as paint or enamel.

        A. the average frost line
        B. footings of the structure
        C. surface of the grade line
        D. permanent moisture level

Timed Exam 5-9   What is the adjustment factor percentage for more than three current-carrying conductors in a raceway or cable if there are 21 current-carrying conductors in a 2 inch EMT.

        A. 35
        B. 40
        C. 45
        D. 70

Timed Exam 5-10   A receptacle installed outdoors in a location protected from the weather or in other damp locations shall have an enclosure for the receptacle that is weatherproof when the receptacle is _____.

        A. being used by a cord plugged into it
        B. open
        C. covered
        D. energized

Timed Exam 5-11   Grounding electrode plates shall be installed not less than _____ in. below the surface of the earth.

        A. 45
        B. 95
        C. 30
        D. 60

Timed Exam 5-12   RMC threadless couplings and connectors used with conduit shall be made tight. Where buried in masonry or concrete, they shall be the ____ type.

        A. watertight
        B. airtight
        C. waterproof
        D. concrete tight

Timed Exam 5-13   Grounding electrode conductor(s) shall be installed in one continuous length without a splice or joint except splicing of the wire-type grounding electrode conductor shall be permitted by irreversible compression-type connectors listed as grounding and bonding equipment or by _____ .

        A. soldering
        B. silver soldering
        C. the exothermic welding process
        D. arc welding

Timed Exam 5-14   An outlet on a 20 amp rated circuit must be rated at _____ amps.

        A. 15
        B. 20
        C. 30
        D. Either A or B

Timed Exam 5-15   A unit load of not less than _____ VA per square foot is required for general lighting in Restaurant occupancies.

        A. 1
        B. 2
        C. 3
        D. 4

Timed Exam 5-16   For other than a totally enclosed switch- board or switchgear, a space not less than _____ shall be provided between the top of the switchboard or switchgear and any combustible ceiling, unless a noncom- bustible shield is provided between the switchboard or switchgear and the ceiling.

        A. 12 inches
        B. 1 ft.
        C. 3 ft
        D. none of the above

Timed Exam 5-17   Conductors shall be considered outside of a building or other structure where they are installed under not less than ____ in. of concrete beneath a building or other structure or where the are installed within a building or other structure in a raceway that is encased in concrete or brick not less than ____ in. thick

        A. 3, 3
        B. 4, 4
        C. 6, 6
        D. 2, 2

Timed Exam 5-18   What is the maximum allowable ampacity for No. 6 aluminum conductors type THHN, 194 F rated insulation, installed in an ambient temperature of 30 C.

        A. 70
        B. 58.2
        C. 60.0
        D. 55

Timed Exam 5-19   LFMC shall be securely fastened in place by an approved means within ____ in. of each box, cabinet, conduit body, or other conduit termination and shall be supported and secured at intervals not to exceed ____ ft. Exceptions ignored.

        A. 8, 4
        B. 12, 4 1/2
        C. 12, 6
        D. 18, 4.5

Timed Exam 5-20   Where a branch circuit supplies continuous loads or any combination of continuous and noncontinuous loads, the minimum branch-circuit conductor size, shall have an allowable ampacity not less than the noncontinuous load plus _____ percent of the continuous load. Exceptions ignored.

        A. 50
        B. 75
        C. 100
        D. 125

Timed Exam 5-21   A 30-ampere branch circuit shall be permitted to supply fixed lighting units with heavy-duty lampholders in other than a dwelling unit(s) or utilization equipment in any occupancy, and the rating of any one cord-and-plug-connected utilization equipment shall not exceed _____ percent of the branch-circuit ampere rating.

        A. 50
        B. 80
        C. 100
        D. 125

Timed Exam 5-22   Enclosures for overcurrent devices shall be mounted in a _____ position unless that is shown to be impracticable.

        A. perpendicular
        B. upright
        C. horizontal
        D. vertical

Timed Exam 5-23   The connection between two or more portions of the equipment grounding conductor.

        A. Aluminum Wire
        B. Branch Circuit
        C. Copper Wire
        D. Equipment Bonding Jumper

Timed Exam 5-24   A building or other structure served shall be supplied by _____ service. Exceptions ignored

        A. only one
        B. only two
        C. only three
        D. only four

Timed Exam 5-25   Raceways shall be provided with expansion fittings where necessary to _____ for thermal expansion and contraction.

        A. provide enhancement
        B. provide restriction
        C. capacitance
        D. compensate

Timed Exam 5-26   Small Appliance Circuit Load. In each dwelling unit, the load shall be calculated at _____ for each 2-wire mall-appliance branch circuit as covered by 210.11(C)(1). Where the load is subdivided through two or more feeders, the calculated load for each shall include not less than _____ for each 2-wire smallappliance branch circuit.

      A. 500 volt-amperes, 1500 volt-amperes
      B. 1000 volt-amperes, 1000 volt-amperes
      C. 1500 volt-amperes, 1500 volt-amperes
      D. 2000 volt-amperes, 1000 volt-amperes

Timed Exam 5-27   In Dwelling units, permanently installed electric baseboard heaters equipped with factory-installed receptacle outlets or outlets provided as a separate assembly by the manufacturer shall be permitted as the required outlet or outlets for the wall space utilized by such permanently installed heaters. Such receptacle outlets shall _____.

      A. be rated the same as the heater circuits
      B. be connected to and controlled by the heater circuits
      C. not be connected to the heater circuits
      D. be install above the heating elements

Timed Exam 5-28   A metal underground water pipe in direct contact with the earth for _____ ft or more, including any metal well casing effectively bonded to the pipe, and electrically continuous, or made electrically continuous by bonding around insulating joints or insulating pipe, to the points of connection of the grounding electrode conductor and the bonding conductor(s) or jumper(s), if installed.

      A. 8
      B. 10
      C. 12
      D. 15

Timed Exam 5-29 In or under airport runways, including adjacent areas where trespassing prohibited what is the minimum burial depth for direct burial cables or conductors.

    A. 6
    B. 12
    C. 18
    D. 24

Timed Exam 5-30 For screw shell devices with attached leads, the conductor attached to the screw shell shall have a _____ finish and the outer finish of the other conductor shall be of a solid color that will not be confused the grounded conductor terminal.

    A. green
    B. black
    C. gold
    D. white or gray

Timed Exam 5-31 A multi-family dwelling has 40 dwelling units. Meters are in two banks of 20 each with individual feeders to each dwelling unit. One-half of the dwelling units are equipped with electric ranges not exceeding 12 kW each. Assume range kW rating equivalent to kVA rating in accordance with 220.55. Other half of ranges are gas ranges. Area of each dwelling unit is 840 sq. ft. Laundry facilities on premises are available to all tenants. Add no circuit to individual dwelling unit. What is the general lighting load for each unit, in VA.

    A. 1520
    B. 2050
    C. 2520
    D. 3630

Timed Exam 5-32 For luminaires near swimming pools, listed low-voltage luminaires not requiring grounding, not exceeding the low- voltage contact limit, and supplied by listed transformers or power supplies that comply with 680.23(A)(2) shall be per- mitted to be located less than _____ from the inside walls of the pool.

    A. 2 ft
    B. 5 ft
    C. 10 ft
    D. 15 ft

Timed Exam 5-33   A 3000 sq. ft. store, has 30 ft of show window. There are a total of 80 duplex receptacles. The service is 120/240 V, single phase 3-wire service. Actual connected lighting load is 8500 VA. What is the total calculated general lighting load in VA.

    A. 6000
    B. 9000
    C. 12000
    D. 16200

Timed Exam 5-34   The required branch circuit copper conductor size for circuit wires of a 40 amp rated circuit is No. _____.

    A. 12
    B. 10
    C. 8
    D. 6

Timed Exam 5-35   The metal frame of the building or structure, where _____ ft or more of a single structural metal member in direct contact with the earth or encased in concrete that is in direct contact with the earth shall be permitted to be used as a grounding electrode.

    A. 5
    B. 8
    C. 10
    D. 12

Timed Exam 5-36   The load for household electric clothes dryers in a dwelling unit(s) shall be either _____ watts or the nameplate rating, whichever is larger, for each dryer served.

    A. 4000
    B. 5000
    C. 7500
    D. 10,000

Timed Exam 5-37   When using Electrical Metallic Tubing: Type EMT wiring systems, when equipment grounding is required, _____

       A. an equipment ground screw connected to an EMT pipe clamp can be used to ground the equipment.
       B. EMT will provide sufficient grounding means.
       C. a separate equipment grounding conductor shall be installed in the conduit.
       D. any of the above

Timed Exam 5-38   The nominal voltage of branch circuits shall not exceed _____ nominal, between conductors In dwelling units and guest rooms or guest suites of hotels, motels, and similar occupancies that supply the terminals of luminaries (lighting fixtures) and cord-and-plug-connected loads 1440 volt-amperes, nominal, or less or less than 1/4hp.

       A. 400
       B. 220
       C. 120
       D. 110

Timed Exam 5-39   Bare _____ grounding electrode conductors shall not be used where in direct contact with masonry or the earth or where subject to corrosive conditions.

       A. aluminum
       B. copper
       C. copper-clad aluminum
       D. aluminum or copper-clad aluminum

Timed Exam 5-40   Snap switches, including dimmer and similar control switches, shall be connected to an equipment grounding conductor and shall provide a means to connect metal faceplates to the equipment grounding conductor, whether or not a metal faceplate is installed. Exception ignored.

       A. This is True to the Code
       B. This is False to the Code

Timed Exam 5-41 All switches and circuit breakers used as switches shall be located so that they may be operated from a readily accessible place and shall be installed such that the center of the grip of the operating handle of the switch or circuit breaker, when in its highest position, is not more than _____ above the floor or working platform. Exceptions ignored.

    A. 5 ft 2 in.
    B. 4 ft 7 in.
    C. 6 ft 6 in.
    D. 6 ft 7 in.

Timed Exam 5-42 What is the demand factor percentage for 9 household electric clothes dryers in a multi-family dwelling unit.

    A. 100
    B. 85
    C. 65
    D. 55

Timed Exam 5-43 The required branch circuit copper conductor size for taps of a 15 amp rated circuit is No. _____.

    A. 18
    B. 14
    C. 12
    D. 10

Timed Exam 5-44 Where the assembly, including the over-current devices protecting the feeder(s), is listed for operation at _____ percent of its rating, the ampere rating of the overcurrent device shall be permitted to be not less than the sum of the continuous load plus the noncontinuous load.

    A. 80
    B. 125
    C. 100
    D. 150

Timed Exam 5-45 What is the adjustment factor percentage for more than three current-carrying conductors in a raceway or cable if there are 4 current-carrying conductors in a 3/4 inch EMT.

       A. 45
       B. 50
       C. 70
       D. 80

Timed Exam 5-46 High-impedance grounded neutral systems in which a grounding impedance _____ ground-fault current to a low value shall be permitted for 3-phase ac systems of 480 volts to 1000 volts where certain conditions are met.

       A. usually a resistor
       B. of not less greater that 10 ohms
       C. of not greater than 10 mA.
       D. of not greater than 240 Volts.

Timed Exam 5-47 Outlet boxes that do not enclose devices or utilization equipment shall have a minimum internal depth of _____.

       A. 3/4
       B. 1
       C. 1/2
       D. 15/16

Timed Exam 5-48 The branch circuit serving emergency lighting and power circuits _____ part of a multiwire branch circuit.

       A. shall be
       B. shall not be
       C. can be
       D. is sometimes

Timed Exam 5-49   Each building or structure disconnect shall _____ disconnect all ungrounded supply conductors it controls and shall have a fault-closing rating not less than the maximum available short-circuit current available at its supply terminals. Exceptions ignored.

    A. automatically
    B. remotely
    C. directly
    D. simultaneously

Timed Exam 5-50   Where a feeder supplies continuous loads or any combination of continuous and noncontinuous loads, the rating of the overcurrent device shall not be less than the noncontinuous load plus _____ percent of the continuous load. Exceptions ignored.

    A. 80
    B. 100
    C. 125
    D. 150

Timed Exam 5-51   An energy management system shall not override the load shedding controls put in place to ensure the minimum electrical capacity. Which of the following systems are not covered under this requirement.

    A. Fire pumps
    B. Emergency systems
    C. Legally required standby systems
    D. Critical operations power systems
    E. none of them

Timed Exam 5-52   Where the assembly, including the overcurrent devices protecting the feeder(s), is listed for operation at _____ percent of its rating, the allowable ampacity of the feeder conductors shall be permitted to be not less than the sum of the continuous load plus the noncontinuous load.

    A. 80
    B. 100
    C. 125
    D. 150

Timed Exam 5-53   Exposed, normally non-current-carrying metal parts of fixed equipment supplied by or enclosing conductors or components that are likely to become energized shall be connected to an equipment grounding conductor under any of the following conditions:

    A. Where within 2.5 m (8 ft) vertically or 1.5 m (5 ft) horizontally of ground or grounded metal objects and subject to contact by persons
    B. Where located in a wet or damp location and not isolated
    C. Where in electrical contact with metal
    D. provided with GFI protectorWhere equipment operates with any terminal at over 150 volts to ground
    E. any of these

Timed Exam 5-54   If required by the authority having jurisdiction, a _____ shall be provided prior to the installation of the feeders.

    A. feeder conductor meter test
    B. diagram showing feeder details
    C. sample calculation
    D. flexible mirror scope inspection device

Timed Exam 5-55   A warehouse storage area uses high pressure sodium lighting. What is the minimum lighting load in the warehouse area if it has 10,500 sq. ft. for storage.

    A. 2625
    B. 5262
    C. 3510
    D. 31500

Timed Exam 5-56   In a service installation for a dwelling what is the maximum service feeder rating for 4/0 aluminum service conductors in amperes.

    A. 150
    B. 175
    C. 200
    D. 225

Timed Exam 5-57   What is the ampacity rating of a two conductor copper No. 8 Type SCE flexible cord rated a 60 C. with an ambient temperature o 30 C.

        A. 60
        B. 55
        C. 48
        D. 65

Timed Exam 5-58   Where oil switches or air, oil, vacuum, or sulfur hexafluoride circuit breakers constitute a building disconnecting means, an isolating switch with _____ and meeting the requirements of 230.204(B), (C), and (D) shall be installed on the supply side of the disconnecting means and all associated equipment. Exception ignored.

        A. visible break contacts
        B. a explosion proof housing
        C. a hand operated lever
        D. non of the above

Timed Exam 5-59   All switchboards, switch-gear, panel-boards, distribution boards, and motor control centers shall be located in _____ spaces and protected from damage.

        A. interior
        B. exterior
        C. dedicated
        D. dry

Timed Exam 5-60   Show windows shall be calculated in accordance with either of the following either the unit load per outlet as required in other provisions of this section or at 200 volt-amperes per _____ ft of show window.

        A. 1
        B. 2
        C. 3
        D. 4

Timed Exam 5-61   A grounding electrode system consisting of a ground ring shall be buried at a depth below the earth's surface of not less than _____ in.

        A. 18
        B. 24
        C. 30
        D. 60

Timed Exam 5-62   Underground service conductors shall have sufficient ampacity to carry the current for the load as calculated and the minimum size of the conductors shall not be smaller than _____ aluminum or copper-clad aluminum.

        A. 6 AWG copper or 6 AWG
        B. 8 AWG copper or 6 AWG
        C. 4 AWG copper or 6 AWG
        D. 8 AWG copper or 8 AWG

Timed Exam 5-63   IMC larger than metric designator trade size _____ shall not be used.

        A. 2
        B. 3
        C. 4
        D. 5

Timed Exam 5-64   Lampholders of the screw-shell type shall be installed for use as lampholders only. Where supplied by a circuit having a grounded conductor, the grounded conductor shall be connected to the _____.

        A. center terminal
        B. green terminal
        C. grounding terminal
        D. screw shell

Timed Exam 5-65 Plate electrodes are permitted to be used for grounding electrodes if each plate electrode is expose not less than _____ sq. ft of surface to exterior soil.

    A. 2
    B. 4
    C. 6
    D. 9

Timed Exam 5-66 Electrical Nonmetallic Tubing (ENT) is not permitted to be used in any building not exceeding _____ floors above grade, unless a fire sprinkler system(s) is installed in accordance with NFPA 13-2002, Standard for the Installation of Sprinkler Systems, on all floors.

    A. one
    B. two
    C. three
    D. five

Timed Exam 5-67 What is the maximum allowable ampacity for No. 6 aluminum conductors type THHN, 194 F rated insulation, installed in an ambient temperature of 30 C. There are 3 current carrying conductors in the cable.

    A. 50
    B. 75
    C. 60
    D. 55

Timed Exam 5-68 XHH insulation has what which characteristic trade name.

    A. Moisture-resistant thermoset
    B. Modified ethylene tetrafluoro-ethylene
    C. Extended polytetra-fluoro-ethylene
    D. Thermoset

Timed Exam 5-69   The feeder conductor ampacity shall not be less than that of the _____ where the feeder conductors carry the total load supplied by service conductors with an ampacity of 55 am-peres or less.

    A. the total load
    B. 4/0
    C. No. 8
    D. service conductors

Timed Exam 5-70   Except as elsewhere required or permitted by the Code, live parts of electrical equipment operating at _____ volts or more shall be guarded against accidental contact by approved enclosures or by any of the following other means.

    A. 35
    B. 50
    C. 80
    D. 100

Timed Exam 5-71   For pools, spas, fountains, and similar locations, lighting systems shall be installed not less than _____ horizontally from the nearest edge of the water, unless permitted by Article 680.

    A. 9 m (30 ft)
    B. 6 m (20 ft)
    C. 12 m (40 ft)
    D. 3 m (10 ft)

Timed Exam 5-72   Where outdoor lampholders are attached as pendants, the connections to the circuit wires shall be _____.

    A. staggered
    B. aligned
    C. taped
    D. sleeved

Timed Exam 5-73   A grounding terminal or grounding-type device can _____.

        A. be used by grounded conductor
        B. bond the grounded conductor to the neutral
        C. not be used for purposes other than grounding
        D. be used in lieu of a GFI

Timed Exam 5-74   At least one 125-volt, single- phase, 15- or 20-ampere-rated receptacle outlet shall be installed within 450 mm (18 in) of the top of a show window for each _____ or major fraction thereof of show window area measured horizontally at its maximum width.

        A. 6 linear ft
        B. 12 linear ft
        C. 10 linear ft
        D. 8 linear ft

Timed Exam 5-75   Where a portable generator, rated 15 kW or less, is installed using a flanged inlet or other cord- and plug-type connection, a disconnecting means shall _____ where un- grounded conductors serve or pass through a building or structure.

        A. not be required
        B. be required
        C. be readily accessible
        D. provided within 6 ft. of

Timed Exam 5-76   The connection of a grounding electrode conductor or bonding jumper to a grounding electrode shall be accessible except, an encased or buried connection in a _____, driven, or buried grounding electrode shall not be required to be accessible.

        A. rigid metallic tube
        B. concrete-encased
        C. flexible metallic tube
        D. electrical metallic tube

Timed Exam 5-77   What is the maximum percent of EMT conduit total interior area that can be filled if more that 2 conductors are used in the tube.

       A. 53
       B. 31
       C. 40
       D. 60

Timed Exam 5-78   A 3000 sq. ft. store , has 30 ft of show window. There are a total of 80 duplex receptacles. The service is 120/240 V, single phase 3-wire service. Actual connected lighting load is 8500 VA. What is the total calculated show window lighting load in VA.

       A. 6000
       B. 9000
       C. 5000
       D. 7000

Timed Exam 5-79   Type AC cable shall not be used _____.

       A. In damp or wet locations
       B. In air voids of masonry block or tile walls where such walls are exposed or subject to excessive moisture or dampness
       C. where subject to physical damage
       D. all the above
       E. none of the above

Timed Exam 5-80   Where the load is calculated on the basis of volt-amperes per square meter or per square foot, the wiring system up to and including the branch-circuit panelboard(s) shall be provided to serve not less than the calculated load. This load shall be _____ among multioutlet branch circuits within the panelboard(s). Branch-circuit overcurrent devices and circuits shall only be required to be installed to serve the connected load.

       A. connected in parallel
       B. concentrated
       C. evenly proportioned
       D. determined and calculated

Section 1 Answers

#	Ans	Answer	Reference
1.	D	continuity	250.12 Clean Surfaces.
2.	A	Yes	Table 250.3 Additional Grounding and Bonding Requirements, Article 650
3.	A	not be required	702.12 Outdoor Generator Sets. (B) Portable Generators 15 kW or Less
4.	D	12,200	Annex D, Example D3 Store Building
5.	A	1	220.14 Other Loads All Occupancies, (G) Show Windows.
6.	D	incandescent luminaries	404.14 Rating and Use of Snap Switches, (E) Dimmer Switches.
7.	D	circular mils	110.6 Conductor Sizes.
8.	C	exothermic welding process	250.64 Grounding Electrode Conductor Installation, (C) Continuous.
9.	A	1	110.26 Spaces About Electrical Equipment, (C) Entrance to and Egress from Working Space., (1) Minimum Required.
10.	A	Operation at a substantially constant load for an indefinitely long time.	ARTICLE 100 Definitions: Duty, Continuous
11.	C	48, 60	680.9 Electric Pool Water Heaters.
12.	A	only one	230.2 Number of Services.
13.	C	shall not be	700.19 Multiwire Branch Circuits.
14.	D	D	250.102 Bonding Conductors and Jumpers, (A) Material.
15.	C	45	310.15: Table 310.15(B)(2)(a) Adjustment Factors for More Than Three Current-Carrying Conductors in a Raceway or Cable
16.	D	distinguishable by color, tagging, or other effective means from those that are part of the fire-rated design	300.11 Securing and Supporting, (A) Secured in Place (1) Fire-Rated Assemblies.
17.	D	3 m (10 ft)	411.5 Specific Location Requirements. (B) Pools, Spas, Fountains, and Similar Locations
18.	A	tapped hole	314.40 Metal Boxes, Conduit Bodies, and Fittings, (D) Grounding Provisions.

19.	A	copper, aluminum, or copper-clad aluminum	250.62 Grounding Electrode Conductor Material.
20.	D	24	300.5: Table 300.5 Minimum Cover Requirements, 0 to 1000 Volts, Nominal, Burial
21.	C	10, 6	440.64 Supply Cords.
22.	B	double	314.16 Number of Conductors in Outlet, Device, and Junction Boxes, and Conduit Bodies. (4) Device or Equipment Fill.
23.	C	Mechanical protection	225.20 Mechanical Protection of Conductors.
24.	A	This is true	250.118 Types of Equipment Grounding Conductors, (9)
25.	C	factory assembled in parallel	240.8 Fuses or Circuit Breakers in Parallel.
26.	D	24	300.5: Table 300.5 Minimum Cover Requirements, 0 to 1000 Volts, Nominal, Burial in Millimeters (Inches)
27.	B	9000	Annex D, Example D3 Store Building
28.	A	usually a resistor	250.36 High-Impedance Grounded Neutral Systems.
29.	A	Ampacity	ARTICLE 100 Definitions
30.	D	125	210.20 Overcurrent Protection. (A) Continuous and Noncontinuous Loads.
31.	C	dedicated	110.26 Spaces About Electrical Equipment (E) Dedicated Equipment Space.
32.	E	none of the above	410.151 Installation, (A) Lighting Track, (C) Locations Not Permitted.
33.	C	Lead sheath	310.13: Table 310.104(A) Conductor Applications and Insulations Rated 600 Volts
34.	D	range, counter-mounted cooking unit, or sink	210.52 Dwelling Unit Receptacle Outlets. (C) (4) Separate Spaces.
35.	D	50	240.5 Protection of Flexible Cords,B. (2) Fixture Wire.
36.	D	Thermoset	300.13: Table 310.104(A) Conductor Applications and Insulations Rated 600 Volts

37.	C	36	314.71 Size of Pull and Junction Boxes, Conduit Bodies, and Handhole Enclosures. (1) Distance to OppositeWall.
38.	A	Calculated Service load 29,200 VA - Minimum Service Rating 122A	Annex D, Example D2(b) Optional Calculation for One-Family Dwelling, Air Conditioning Larger Than Heating
39.	D	30	250.53 Grounding Electrode System Installation, (G) Rod and Pipe Electrodes.
40.	C	flexible cord to facilitate the removal or disconnection for maintenance or repair	680.7 Cord-and-Plug-Connected Equipment.
41.	C	3	ARTICLE 100 Definitions, Continuous Load
42.	C	the applied voltage	230.30 Insulation.
43.	D	be guarded or isolated	240.41 Arcing or Suddenly Moving Parts (B)
44.	A	1/8, 125	220.18 Maximum Loads, (A) Motor-Operated and Combination Loads.
45.	C	18	300.5: Table 300.5 Minimum Cover Requirements, 0 to 1000 Volts, Nominal, Burial
46.	C	unless installed as part of a listed tub or shower assembly	404.4 Damp or Wet Locations (C) Switches in Tub or Shower Spaces
47.	D	145	Table 310.60(C)(69) Ampacities of Insulated Single Copper Conductor Isolated in Air Based
48.	C	be permitted to serve as the disconnecting means	422.31 Disconnection of Permanently Connected Appliances, (A) Rated at Not Over 300 Volt-Amperes or 1/8Horsepower.
49.	C	For system voltages of 1000 volts maximum	360.10 Uses Permitted. FMT
50.	C	100	220.42: Table 220.42 Lighting Load Demand Factors
51.	A	grounded	240.23 Change in Size of Grounded Conductor.
52.	C	29,200 VA, 122 A	Annex D, Example D2(b) Optional Calculation for One-Family Dwelling, Air Conditioning Larger Than Heating
53.	C	not be used	225.26 Vegetation as Support.
54.	B	18 in , 24 in	625.50 Location.

55.	D	closed in an approved manner	312.5 Cabinets, Cutout Boxes, and Meter Socket Enclosures. (A) Openings to Be Closed.
56.	B	final grade	230.24 Clearances. (B) Vertical Clearance for Overhead Service Conductors
57.	D	not be used	110.13 Mounting and Cooling of Equipment. (A) Mounting.
58.	C	covered	406.9 Receptacles in Damp or Wet Locations. (A) Damp Locations.
59.	D	80	310.15: Table 310.15(B)(2)(a) Adjustment Factors for More Than Three Current-Carrying Conductors in a Raceway or Cable
60.	B	ungrounded conductors	422.47 Water Heater Controls.
61.	C	25 ft	625.16 Means of Coupling. (C) Overall Cord and Cable Length.
62.	D	four	220.53 Appliance Load Dwelling Unit(s).
63.	C	Moisture- and heat-resistant thermoplastic	300.13: Table 310.104(A) Conductor Applications and Insulations Rated 600 Volts
64.	A	5	680.22 Area Lighting, Receptacles, and Equipment. (C) Switching Devices.
65.	A	This is True to the Code	404.9 Provisions for General-Use Snap Switches (B) Grounding.
66.	D	wall switch controlled lighting outlet	210.70 Lighting Outlets Required. (A) (1) Habitable Rooms.
67.	A	Conduit Body	ARTICLE 100 Definitions, Conduit Body
68.	B	2	210.52 Dwelling Unit Receptacle Outlets, (A) General Provisions. (2) Wall Space. (1)
69.	A	10,125 VA 40500x.25=10,125	220.12: Table 220.12 General Lighting Loads by Occupancy,
70.	C	five	334.24 Bending Radius.
71.	D	Neutral Point.	ARTICLE 100 - DEFINITIONS
72.	A	dry walls	368.10 Uses Permitted, (C) (1) Walls.
73.	A	Luminaire	ARTICLE 100 -DEFINITIONS

74.	A	0	300.5: Table 300.5 Minimum Cover Requirements, 0 to 1000 Volts, Nominal, Burial in Millimeters (Inches)
75.	B	40	374.5 Maximum Number of Conductors in Raceway
76.	C	An unintentional, electrically conducting connection between an ungrounded conductor of an electrical circuit and the normally non-current-carrying conductors, metallic enclosures, metallic raceways, metallic equipment, or earth.	ARTICLE 100 Definitions: Ground Fault
77.	C	2520	Annex D, Example D4(a) Multifamily Dwelling
78.	C	Between a dwelling and separate garage unit	340.12 Uses Not Permitted.
79.	D	grounded	230.22 Insulation or Covering. Exception:
80.	C	the total ampere ratings of such units and not on the total watts of the lamps	220.18 Maximum Loads, (B) Inductive and LED Lighting Loads.
81.	D	Practical Safeguarding.	ARTICLE 90 Introduction
82.	C	125	645.5 Supply Circuits and Interconnecting Cables, (A) Branch-Circuit Conductors.
83.	D	2, 2	230.6 Conductors Considered Outside the Building.
84.	A	6000	Annex D, Example D3 Store Building
85.	A	.412	Chapter 9 Tables: Table 5 Dimensions of Insulated Conductors and Fixture Wires
86.	A	25 ft	210.63 Heating, Air-Conditioning, and Refrigeration Equipment Outlet.
87.	A	14	210.24: Table 210.24 Summary of Branch-Circuit Requirements
88.	B	the ceiling assembly, including the ceiling support wires	300.11 Securing and Supporting, (A) Secured in Place (1) Fire-Rated Assemblies.
89.	A	8	368.12 Busways, Uses Not Permitted. (E) Working Platform.

90.	C	two or more 20-ampere	210.52 Dwelling Unit Receptacle Outlets (B) Small Appliances., (1) Receptacle Outlets Served.
91.	D	A and B	250.52 Grounding Electrodes, (B) Not Permitted for Use as Grounding Electrodes..
92.	C	permitted as the sole support	300.11 Securing and Supporting, (A) Secured in Place.
93.	D	6	210.52 Dwelling Unit Receptacle Outlets. (A) General Provisions. (1) Spacing.
94.	A	Where protection of the contained conductors is required from vapors, liquids, or solids	356.10 Uses Permitted. (2)
95.	B	55	400.5: Table 400.5(A)(2) Ampacity of Cable Types SC, SCE, SCT, PPE, G, G-GC, and W.
96.	C	40	Article 348 Flexible Metal Conduit (FMC) and Chapter 9, Table 1
97.	A	CO/ALR	406.3 Receptacle Rating and Type, (C) Receptacles for Aluminum Conductors.
98.	A	1 VA	220.12: Table 220.12 General Lighting Loads by Occupancy
99.	C	2	250.50 Grounding Electrode System, (4) Ground Ring.
100.	B	Intersystem Bonding Termination.	ARTICLE 100 -DEFINITIONS
101.	B	12	Table 344.30(B)(2) Supports for Rigid Metal Conduit
102.	C	16	Annex C: Table C.1 Maximum Number of Conductors or FixtureWires in Electrical Metallic Tubing (EMT) (Based on Table 1, Chapter 9)
103.	B	untrained persons	90.1 Purpose.(C) Relation to Other International Standards.
104.	B	concrete-encased	250.68 Grounding Electrode Conductor and Bonding Jumper Connection to Grounding Electrodes. (A) Accessibility. Exception No. 1
105.	A	In storage battery rooms	348.12 Uses Not Permitted
106.	B	Service Drop	ARTICLE 100 Definitions

107.	D	B and C	220.84 Multifamily Dwelling, (A) Feeder or Service Load.
108.	A	3, 8	410.10 Luminaries (Fixtures) in Specific Locations, (D) Bathtub and Shower Areas.
109.	B	125	422.10 Branch-Circuit Rating, 422.10 Branch-Circuit Rating. (A) Individual Circuits.
110.	C	125	215.3 Overcurrent Protection.
111.	B	conform with 490.24.	110.36 Circuit Conductors.
112.	D	1/2	314.24 Depth of Boxes. (A) Outlet Boxes Without Enclosed Devices or Utilization Equipment.
113.	B	dry location	680.23 Underwater Luminaries (Lighting Fixtures). (B) Wet-Niche Luminaires. (6) Servicing.
114.	D	A or B	250.119 Identification of Equipment Grounding Conductors. (C) Flexible Cord.
115.	B	Disconnecting Means	ARTICLE 100 Definitions
116.	C	Voltage (of a circuit).	ARTICLE 100 Definitions
117.	D	screw shell	410.90 Screw-Shell Type.
118.	A	a lower	225.40 Access to Overcurrent Protective Devices.
119.	D	open or closed	225.38 Disconnect Construction. (D) Indicating.
120.	C	In a two story building	362.10 Uses Permitted
121.	C	support	362.12 Uses Not Permitted.
122.	D	Bonding jumpers	250.92 Services,
123.	C	30	250.53 Grounding Electrode System Installation, (F) Ground Ring.
124.	B	8 AWG copper or 6 AWG	230.31 Size and Rating. (B) Minimum Size.
125.	B	5000	220.54 Electric Clothes Dryers Dwelling Unit(s).
126.	C	arc-fault circuit interrupter, combination-type,	210.12 Arc-Fault Circuit-Interrupter Protection. (A) Dwelling Units.
127.	B	15	225.39 Rating of Disconnect (A) One-Circuit Installation.
128.	C	serve more than one kitchen	210.52 Dwelling Unit Receptacle Outlets. (3) Kitchen Receptacle Requirements.
129.	B	ungrounded	225.31 Disconnecting Means.
130.	A	Bonding Conductor or Jumper	ARTICLE 100 Definitions

131.	C	30, 90	110.26 Spaces, (2) Width of Working Space.
132.	C	300	404.8 Accessibility and Grouping, (B) Voltage Between Adjacent Devices.
133.	A	6	300.5: Table 300.5 Minimum Cover Requirements, 0 to 1000 Volts, Nominal, Burial
134.	C	200	310.15: Table 310.15(B)(6) Conductor Types and Sizes for 120/240-Volt, 3-Wire, Single-Phase Dwelling Services and Feeders. Conductor Types RHH, RHW, RHW-2, THHN, THHW, THW, THW-2, THWN, THWN-2, XHHW, XHHW-2, SE, USE, USE-2
135.	D	Luminaire	ARTICLE 100 Definitions, Luminaire
136.	D	Inverse Time	ARTICLE 100 Definitions
137.	C	as near as practicable to, and preferably in the same area as,	250.30 Grounding Separately Derived Alternating- Current Systems. (A) Grounded Systems. (4) Grounding Electrode.
138.	A	inadvertent	95.4 Continuity of Power.
139.	C	.25, 2	250.50 Grounding Electrode System, (7) Plate Electrodes.
140.	A	water meters or filtering devices and similar equipment	250.53 Grounding Electrode System Installation, (D) Metal Underground Water Pipe.(1) Continuity.
141.	A	or project therefrom	312.3 Position in Wall.
142.	B	accessible	ARTICLE 100 Definitions, Accessible (as applied to equipment).
143.	C	Effective Ground-Fault Current Path.	250.4 General Requirements for Grounding and Bond- ing. (A) Grounded Systems. (5) Effective Ground-Fault Current Path.
144.	A	calculated	210.11 Branch Circuits Required. (A) Number of Branch Circuits.
145.	B	RNC	250.118 Types of Equipment Grounding Conductors.
146.	E	all of the above	680.10 Underground Wiring Location.
147.	A	secured	312.5 Cabinets, Cutout Boxes, and Meter Socket Enclosures (C) Cables.

148.	B	damp	406.9 Receptacles in Damp or Wet Locations. (A) Damp Locations.
149.	C	1 1/2	300.1 Scope, (C) Metric Designators and Trade Sizes.
150.	B	Permanent ladders or stairways	110.32 Work Space About Equipment. (B) Access.
151.	D	Fire-Resistive Cable System.	728.2 Definition. Fire-Resistive Cable System.
152.	A	Yes	Annex D, Example D4(a) Multifamily Dwelling
153.	A	locked or sealed	230.93 Protection of Specific Circuits.
154.	B	removal of a trim or cover	240.83 Marking, (A) Durable and Visible.
155.	D	Cable Management System (Electric Vehicle Supply Equipment)	625.2 Definitions.
156.	D	A & B	600.4 Markings, (A) Signs and Outline Lighting Systems.
157.	A	staggered	225.24 Outdoor Lampholders.
158.	A	8, 3	225.19 Clearances from Buildings for Conductors of Not Over 1000 Volts, Nominal, (A) Above Roofs
159.	A	angle	348.42 Couplings and Connectors.
160.	B	General-Use Switch	ARTICLE 100 Definitions
161.	C	20.25	Table 314.16(A) Metal Boxes 1 for all grounding wires, 1 for fixture stud, 1 for all clamp inside of box = 9X2.25 = 20.25
162.	D	A reliable conductor to ensure the required electrical conductivity between metal parts required to be electrically connected.	ARTICLE 100 Definitions : Bonding Conductor or Jumper.
163.	C	Rainproof	ARTICLE 100 Definitions
164.	B	125 percent	B) Overcurrent Device Ratings.
165.	B	6	250.53 Grounding Electrode System Installation, (B) Electrode Spacing.
166.	B	The installation is in violation of Code requirements because the height requirement is not met., must be 17 feet minimum.	225.60 Clearances over Roadways, Walkways, Rail, Water, and Open Land, Table 225.61 Clearances over Buildings and Other Structures

167.	D	A device intended for the protection of personnel that functions to deenergize a circuit or portion thereof within an established period of time when a current to ground exceeds the values established for a Class A device.	ARTICLE 100 Definitions; Ground-Fault Circuit Interrupter (GFCI).
168.	C	continuous	300.13 Mechanical and Electrical Continuity Conductors, (A) General.
169.	C	Busway	368.2 Definition.
170.	D	vertical	240.33 Vertical Position.
171.	A	not less than 100% of	210.21 (1) Outlet Devices. Outlet devices shall have an ampere rating that is not less than the load to be served and shall comply with 210.21(A) and (B).
172.	C	where exposed to direct sunlight	328.12 Uses Not Permitted.
173.	B	diagram showing feeder details	215.5 Diagrams of Feeders.
174.	C	11.25	Table 310.15(B)(16) Allowable Ampacities of Insulated Conductors 25A X .45 (adjustment factor)= 11.25
175.	B	the required number of poles	404.11 Circuit Breakers as Switches.
176.	B	in parallel in each raceway or cable	250.122 Size of Equipment Grounding Conductors, (F) Conductors in Parallel.
177.	C	insulated	250.119 Identification of Equipment Grounding Conductors.
178.	C	shall be counted once	314.16 Number of Conductors in Outlet, Device, and Junction Boxes, and Conduit Bodies. (1) Conductor Fill
179.	A	three, 24	310.15 Ampacities for Conductors Rated 0-2000 Volts. (A) General (3) Adjustment Factors.
180.	A	100	310.15: Table 310.15(B)(6) Conductor Types and Sizes for 120/240-Volt, 3-Wire, Single-Phase Dwelling Services and Feeders. Conductor Types RHH, RHW, RHW-2, THHN, THHW, THW, THW-2, THWN, THWN-2, XHHW, XHHW-2, SE, USE, USE-2

181.	B	4.5	Chapter 9 Tables, Table 2 Radius of Conduit and Tubing Bends
182.	A	9 sq. in.	680.26 Equipotential Bonding. (C) Pool Water.
183.	D	simultaneously	225.52 Disconnecting Means.(B) Type.
184.	A	located or shielded	240.41 Arcing or Suddenly Moving Parts (A)
185.	D	firestopped	300.21 Spread of Fire or Products of Combustion.
186.	A	bonding of electrical equipment	250.4 General Requirements for Grounding and Bond- ing. (A) Grounded Systems. (3) Bonding of Electrical Equipment.
187.	D	permanent moisture level	250.53 Grounding Electrode System Installation, (A) Rod, Pipe, and Plate Electrodes. 1.
188.	C	grounding electrode conductor	250.24 Grounding Service-Supplied Alternating Current Systems. (A) System Grounding Connections.
189.	D	20, 2	250.52 Grounding Electrodes. (A) Electrodes Permitted for Grounding. (3) Concrete-Encased Electrode.
190.	C	combustible material	110.18 Arcing Parts.
191.	A	3	230.9 Clearances on Buildings, (A) Clearances.
192.	C	electrically operated by a readily accessible, remote-control device in a separate building or structure.	225.52 Disconnecting Means. (A) Location.
193.	D	the same	250.58 Common Grounding Electrode.
194.	D	twice	314.16 Number of Conductors in Outlet, Device, and Junction Boxes, and Conduit Bodies. (1) Conductor Fill
195.	C	Flexible Metal Conduit (FMC)	348.2 Definition.
196.	C	Cabinet	ARTICLE 100 Definitions, Cabinet.
197.	A	Three 15-A, 2-wire or two 20-A, 2-wire circuits	Annex D, Example D1(a) One-Family Dwelling
198.	C	1500 volt-amperes	220.52 (A) Small Appliance and Laundry Loads Dwelling Unit.
199.	D	Service Point	ARTICLE 100 Definitions
200.	C	8	210.24: Table 210.24 Summary of Branch-Circuit Requirements

201.	D	126	Annex B: Table B.310.15(B)(2)(3) Ampacities of Multiconductor Cables with Not More Than Three Insulated Conductors, Rated 0 Through 2000 Volts, in Free Air Based on Ambient Air Temperature of 40°C (104°F) (for Types TC, MC, MI, UF, and USE Cables
202.	D	Round boxes	314.2 Round Boxes.
203.	C	14.5 feet	ARTICLE 680, SWIMMING POOLS, FOUNTAINS, AND SIMILAR INSTALLATIONS, Table 680.8 (A) Overhead Conductor Clearances
204.	B	10	250.52 Grounding Electrode System, (A) Electrodes Permitted for Grounding, (1) Metal Underground Water Pipe.
205.	A	40	225.6 Conductor Size and Support, (B) Festoon Lighting.
206.	C	4	342.20 Size Intermediate Metal Conduit: Type IMC (B) Maximum,
207.	D	Either A or B	210.21: Table 210.21(B)(2) Maximum Cord-and-Plug-Connected Load to Receptacle
208.	C	conductors	300.3 Conductors., (B) Conductors of the Same Circuit.
209.	D	any of the above	250.62 Grounding Electrode Conductor Material.
210.	B	sealed	225.27 Raceway Seal.
211.	A	service-entrance	230.44 Cable Trays.
212.	D	all of the above makes this statement true	225.52 Disconnecting Means (F) Identification.
213.	A	at the point where the conductors receive their supply	240.92 Location in Circuit. A.
214.	A	a bushing or adapter	352.46 Bushings.
215.	B	simultaneously disconnect	210.7 Branch Circuit Receptacle Requirements: Multiple Branch Circuits.
216.	B	white	200.9 Means of Identification of Terminals.
217.	B	100	215.2 Minimum Rating and Size, (A) Feeders Not More Than 600 Volts. Exception: No. 1

218.	A	Location, Wet.	ARTICLE 100 Definitions, Location, Wet.
219.	A	continuous	300.12 Mechanical Continuity Raceways and Cables.
220.	D	18	225.18 Clearance for Overhead Conductors and Cables. (4)
221.	B	For direct burial where listed and marked for the purpose	350.10 (3) Uses Not Permited
222.	C	10	250.52 Grounding Electrodes, (A) Electrodes Permitted for Grounding.(2) Metal Frame of the Building or Structure. (1)
223.	C	grouping all phase conductors together	300.20 Induced Currents in Ferrous Metal Enclosures or Ferrous Metal Raceways. (A) Conductors Grouped Together.
224.	C	equipment bonding jumpers	250.98 Bonding Loosely Jointed Metal Raceways.
225.	B	45	250.53 Grounding Electrode System Installation, (G) Rod and Pipe Electrodes.
226.	C	6	210.50 General. Receptacle outlets shall be installed as specified in 210.52 through 210.63. (C) (C) Appliance Receptacle Outlets.
227.	A	ungrounded	210.10 Ungrounded Conductors Tapped from Grounded Systems.
228.	C	42	220.84: Table 220.84 Optional Calculations Demand Factors for Three or More Multifamily Dwelling Units
229.	D	55	Table 310.15(B)(16) Allowable Ampacities of Insulated Conductors
230.	B	50	110.27 Guarding of Live Parts. A.
231.	D	Equipment Bonding Jumper	ARTICLE 100 Definitions, Bonding Conductor or Jumper
232.	C	bathrooms	230.70 (2) Bathrooms.
233.	D	either inside or outside	225.32 Location.
234.	A	6	300.5: Table 300.5 Minimum Cover Requirements, 0 to 1000 Volts, Nominal, Burial
235.	E	All of the above	690.5 Ground-Fault Protection., (A) Ground-Fault Detection and Interruption.

236.	D	none of the above	695.6 Power Wiring. (D) Pump Wiring.
237.	B	Capable of being operated without exposing the operator to contact with live parts.	ARTICLE 100 Definitions: Externally Operable
238.	D	compensate	300.7 Raceways Exposed to Different Temperatures, (B) Expansion Fittings.
239.	C	18,600	Annex D, Example D1(a) One-Family Dwelling
240.	C	DC-to-DC Converter.	690.2 Definitions., DC-to-DC Converter.
241.	A	material of the conductor	110.14 Electrical Connections.
242.	A	8, 3	300.14 Length of Free Conductors at Outlets, Junctions, and Switch Points.
243.	B	6	680.22 Lighting, Receptacles, and Equipment, (A) Receptacles. (3) Other Receptacles, Location.
244.	C	30	600.5 Branch Circuits. (B) Rating (1) Neon Signs
245.	B	8.5 inches	Table 312.6(B) Minimum Wire-Bending Space at Terminals
246.	C	75	366.56 Splices and Taps, (A) Within Gutters.
247.	C	Installations of communications equipment under the exclusive control of communications utilities located outdoors or in building spaces used exclusively for such installations	90.2 Scope (B) Not Covered. (4)
248.	C	Interactive System	ARTICLE 100 Definitions, Interactive System
249.	B	first installed on-site	225.56 Inspections and Tests. (A) Pre-Energization and Operating Tests.
250.	E	all are acceptable	230.50 Protection Against Physical Damage. (B) All Other Service-Entrance Conductors.
251.	D	general lighting	220.52 Small Appliance and Laundry Loads-Dwelling Unit. (B) Laundry Circuit Load.

252.	C	12 in., 24 in.	210.52 Dwelling Unit Receptacle Outlets. (C) Countertops (1) Wall Counter Spaces.
253.	A	copper, aluminum, or copper-clad aluminum	250.118 Types of Equipment Grounding Conductors.
254.	D	all these outlets require GFI protection	210.8 Ground-Fault Circuit-Interrupter Protection for Personnel. (6) Kitchens
255.	C	cord-and-plug-connected portable	210.19 Branch-Circuit Ratings Conductors Minimum Ampacity and Size. (2) Branch Circuits with More than One Receptacle.
256.	D	Within a building	353.12 Uses Not Permitted.
257.	C	three	362.10 Uses Permitted.
258.	B	lockable in accordance with 110.25.	225.52 Disconnecting Means. (C) Locking.
259.	C	a separate equipment grounding conductor shall be installed in the conduit.	Article 358 Electrical Metallic Tubing: Type EMT, 356.60 Grounding.
260.	C	feeder and branch-circuit disconnect location	225.37 Identification.
261.	A	Suitable	300.31 Covers Required.
262.	C	Fixture Wiring	402.3: Table 402.3 Fixture Wires
263.	D	independently	240.22 Grounded Conductor.
264.	B	100	Annex D, Example D1(a) One-Family Dwelling
265.	C	convection principles	110.13 Mounting and Cooling of Equipment. (B) Cooling.
266.	B	electrical continuity	250.96 Bonding Other Enclosures, (A) General.
267.	D	the observation of a qualified person at all times	110.34 Work Space and Guarding, (C) Locked Rooms or Enclosures.
268.	C	not be connected to the heater circuits	210.52 Dwelling Unit Receptacle Outlets.
269.	C	6, 8	230.202 Service-Entrance Conductors. (A) Conductor Size.
270.	D	55	Table 310.15(B)(16) Allowable Ampacities of Insulated Conductors
271.	C	corrosive	250.62 Grounding Electrode Conductor Material.
272.	B	12, 4 1/2	350.30 Securing and Supporting. (A) Securely Fastened.
273.	D	entrance	600.5 Branch Circuits, (A) Required Branch Circuit.
274.	A	visible break contacts	225.51 Isolating Switches.

275.	C	all the above	320.12 Uses Not Permitted.
276.	C	8	250.50 Grounding Electrode System, (5) Rod and Pipe Electrodes.
277.	C	24 in. , 7 ft , 18 in.	760.24 Mechanical Execution of Work. (B) Circuit Integrity (CI) Cable.
278.	A	2625, 10500 x .25 = 2625	220.12: Table 220.12 General Lighting Loads by Occupancy
279.	E	all of the above	680.22 Area Lighting, Receptacles, and Equipment, (A) Receptacles, (2) Circulation and Sanitation System, Location.
280.	B	5 ft	680.22 Lighting, Receptacles, and Equipment.(B) Luminaires, Lighting Outlets, and Ceiling-Suspended (Paddle) Fans.(6) Low-Voltage Luminaires.
281.	C	3-wire feeders	215.4 (A) Feeders with Common Neutral Conductor.
282.	C	90	110.40 Temperature Limitations at Terminations.
283.	B	17	402.5: Table 402.5 Allowable Ampacity for Fixture Wires
284.	C	grounded	404.2 Switch Connections, (B) Grounded Conductors.
285.	C	30	250.53 Grounding Electrode System Installation, (H) Plate Electrode.
286.	D	35,000	328.10 Uses Permitted, Medium Voltage Cable,
287.	A	removed	250.96 Bonding Other Enclosures, (A) General.
288.	C	High pressure sodium parking lot lighting	210.2 Table 210.2: Specific-Purpose Branch Circuits
289.	E	any of these	250.110 Equipment Fastened in Place (Fixed) or Connected by Permanent Wiring Methods
290.	D	grounding electrode system	250.106 Lightning Protection Systems.
291.	C	Orange	110.15 High-Leg Marking.
292.	C	3 ft.	408.18 Clearances, (A) From Ceiling.
293.	A	Yes	210.52: Figure 210.52(C)(1) Determination of Area Behind Sink or Range.

294.	B	Bonding Jumper	ARTICLE 100 Definitions, Bonding Conductor or Jumper.
295.	D	Article 682	Table 250.3 Additional Grounding and Bonding Requirements, Article 682
296.	C	connected to earth	250.4 General Requirements for Grounding and Bonding. (A) Grounded Systems (1) Grounding of Electrical Equipment.
297.	B	Remainder over 50,000	220.42: Table 220.42 Lighting Load Demand Factors
298.	B	14	210.24: Table 210.24 Summary of Branch-Circuit Requirements
299.	C	6	300.14 Length of Free Conductors at Outlets, Junctions, and Switch Points.
300.	A	independent of the building	230.29 Supports over Buildings.
301.	D	all of the above	210.21: Table 210.21(B)(3) Receptacle Ratings for Various Size Circuits
302.	A	above the tunnel floor	110.51 General. (C) Protection Against Physical Damage.
303.	C	15	225.18 Clearance from Ground. (3)
304.	B	35	Table C.1 Maximum Number of Conductors or FixtureWires in Electrical Metallic Tubing (EMT) (Based on Table 1, Chapter 9)
305.	B	208	424.35 Marking of Heating Cables.
306.	B	under a building	300.5 Underground Installations (C) Underground Cables Under Buildings.
307.	D	raceway	300.15 Boxes, Conduit Bodies, or Fittings (J) Luminaries (Fixtures).
308.	D	Both A and C	210.8 Ground-Fault Circuit-Interrupter Protection for Personnel. (A) Dwelling Units.
309.	C	Tap Conductors	240.2 Definitions.
310.	A	copper	110.5 Conductors.
311.	B	Cutout	ARTICLE 100 Definitions
312.	A	2	250.50 Grounding Electrode System, (A) Electrodes Permitted for Grounding, (7) Plate Electrodes.
313.	B	42, six 7-A units 6x7=42	Annex D, Example D2(b) Optional Calculation for One-Family Dwelling, Air Conditioning Larger Than Heating

314.	B	twice	220.55 Electric Ranges and Other Cooking
315.	C	15	Table C.1 Maximum Number of Conductors or FixtureWires in Electrical Metallic Tubing (EMT) (Based on Table 1, Chapter 9)
316.	A	no wiring systems of any type	300.22 Wiring in Ducts, Plenums, and Other Air-Handling Spaces, (A) Ducts for Dust, Loose Stock, or Vapor Removal.
317.	C	3	220.12: Table 220.12 General Lighting Loads by Occupancy
318.	C	4500 VA	Annex D, Example D1(a) One-Family Dwelling
319.	B	Type NMC	334.2 Definitions.
320.	C	the sum of the nameplate ratings of the transformers supplied when only transformers are supplied	215.2 Minimum Rating and Size, (B) (1) Feeders Supplying Transformers.
321.	C	Type NM cable	230.43 Wiring Methods for 1000 Volts, Nominal, or Less.
322.	B	ungrounded conductors	215.7 Ungrounded Conductors Tapped from Grounded Systems.
323.	B	B	210.23 Permissible Loads. (B) 30-Ampere Branch Circuits.
324.	C	rms	620.13 Feeder and Branch-Circuit Conductors. (A) Conductors Supplying Single Motor.
325.	B	5 ft	Solar Photovoltaic (PV) Storage Batteries 690.71 Installation. (H) Disconnects and Overcurrent Protection.
326.	D	white or gray	200.10 Means of Identification of Terminals. (D) Screw Shell Devices with Leads.
327.	A	10	225.18 Clearance from Ground. (1)
328.	D	125	210.19 Branch-Circuit Ratings Conductors Minimum Ampacity and Size. (a)
329.	B	Multiple Fuse	ARTICLE 100 Definitions
330.	D	55	220.54: Table 220.54 Demand Factors for Household Electric Clothes Dryers
331.	C	Connector, Pressure (Solderless)	ARTICLE 100 Definitions, Connector, Pressure (Solderless)

332.	C	100	225.39 Rating of Disconnect. (C) One-Family Dwelling.
333.	B	5	680.10 Underground Wiring Location.
334.	A	Bypass Isolation Switch	ARTICLE 100 Definitions
335.	E	none of them	Article 750 Energy Management Systems, 750.30 Load Management, (A) Load Shedding Controls.
336.	C	At the time of installation, by a distinctive white or gray marking at its terminations. This marking shall encircle the conductor or insulation.	200.6 Means of Identifying Grounded Conductors. (B) Sizes Larger Than 6 AWG. (4)
337.	A	bonded together	250.92 Services. (A) Bonding of Equipment for Services
338.	C	In commercial garages, other than for supplying ceiling outlets or extensions to the area below the floor but not above	372.4 Uses Not Permitted. (3)
339.	D	15,510, Heat Pump and Supplementary Heat 240 V × 24 A = 5760 VA 15 kW Electric Heat, 5760 VA + (15,000 VA × 65%) = 5.76 kVA + 9.75 kVA = 15.51 kVA, if supplementary heat is not on at same time as heat pump, heat pump kVA need not be added to total.	Annex D, Example D2(c) Optional Calculation for One-Family Dwelling with Heat Pump (Single-Phase, 240/120-Volt Service) (see 220.82)
340.	D	securely fastened to the construction	250.64 Grounding Electrode Conductor Installation. (B) Securing and Protection Against Physical Damage.
341.	D	18, 36	422.16 Flexible Cords. (B)(2) Specific Appliances.
342.	A	Class 2 power source	605.6 Lighting Accessories, B) Connection. (4)
343.	D	6 ft 7 in.	404.8 Accessibility and Grouping, (A) Location.

344.	C	heavy-duty type	210.21 Outlet Devices. Outlet devices shall have an ampere rating that is not less than the load to be served and shall comply with 210.21(A) and (B).
345.	A	without the use of splice boxes	300.5 Underground Installations (E) Splices and Taps.
346.	D	exposed and concealed	358.10 Uses Permitted, (A) Exposed and Concealed.
347.	C	automotive vehicles other than mobile homes and recreational vehicles	90.2 Scope. B. Not Covered. (1)
348.	A	of any shape	250.118 Types of Equipment Grounding Conductors. (1)
349.	C	outlets not exceeding 240 volts	210.8 Ground-Fault Circuit-Interrupter Protection for Personnel. (C) Boat Hoists.
350.	C	538 sq ft	694.23 Turbine Shutdown. (A) Manual Shutdown.
351.	D	nonabsorbent	230.27 Means of Attachment.
352.	A	grounded	225.38 Disconnect Construction. Disconnecting, (C) Disconnection of Grounded Conductor.
353.	B	15	406.3 Receptacle Rating and Type, (B) Rating.
354.	D	3, 5	342.30 (A) Securely Fastened. (1) & (2)
355.	C	13	Table 314.16(A) Metal Boxes
356.	D	braces or guys	230.28 Service Masts as Supports. (A)
357.	B	1000	110.54 Bonding and Equipment Grounding Conductors. (A) Grounded and Bonded.
358.	C	1.25, .0625	300.4 Protection Against Physical Damage. (A) Cables and Raceways Through Wood Members. (1) Bored Holes.
359.	A	only one feeder or branch circuit	225.30 Number of Supplies.
360.	D	Type UF	340.2 Definition.
361.	B	18	250.64 Grounding Electrode Conductor Installation, (A) Aluminum or Copper-Clad Aluminum Conductors.

362.	A	ungrounded	230.209 Surge Arresters (Lightning Arresters).
363.	B	2	220.12: Table 220.12 General Lighting Loads by Occupancy
364.	A	opposite	366.56 Splices and Taps, (B) Bare Conductors.
365.	C	9.5	Chapter 9 Tables, Table 2 Radius of Conduit and Tubing Bends
366.	C	115	Table 310.15(B)(16) Allowable Ampacities of Insulated Conductors
367.	D	suitable for use	225.22 Raceways on Exterior Surfaces of Buildings or Other Structures.
368.	B	12	225.18 Clearance from Ground. (2)
369.	B	panelboard	210.4 Multiwire Branch Circuits. (A) General.
370.	A	shall be permitted to be connected in parallel	310.10 (H) Conductors in Parallel. (1) General.
371.	B	guards	110.27 Guarding of Live Parts. (B) Prevent Physical Damage.
372.	B	reduced	358.24 Bends, How Made.
373.	C	environment	300.6 Protection Against Corrosion and Deterioration.
374.	C	100 percent	(D) Selectivity.
375.	A	six, six	225.33 Maximum Number of Disconnects. (A) General.
376.	B	12 linear ft	210.62 Show Windows.
377.	C	a single grounding electrode system.	250.53 Grounding Electrode System Installation, (B) Electrode Spacing.
378.	B	crowding	312.7 Space in Enclosures.
379.	B	10	230.26 Point of Attachment.
380.	D	sealed or plugged at either or both ends	300.5 Underground Installations. (G) Raceway Seals.
381.	D	aluminum or copper-clad aluminum	250.64 Grounding Electrode Conductor Installation, (A) Aluminum or Copper-Clad Aluminum Conductors.
382.	A	white or gray	200.6 Means of Identifying Grounded Conductors. (C) Flexible Cords.
383.	D	ampacity	ARTICLE 100 Definitions. General Ampacity
384.	D	service conductors	215.2 Minimum Rating and Size, (A) (3) Ampacity Relative to Service Conductors
385.	A	either C or D	225.25 Location of Outdoor Lamps.

386.	C	flexible cords, flexible cables, and fixture wires	240.4 Protection of Conductors.
387.	B	Plenum.	ARTICLE 100 Definitions
388.	B	5	366.30 Securing and Supporting, (A) Sheet Metallic Auxiliary Gutters.
389.	A	3 ft	210.52 Dwelling Unit Receptacle Outlets. (D) Bathrooms.
390.	C	nominal voltage	110.4 Voltages.
391.	A	voltage	110.9 Interrupting Rating.
392.	B	ground faults	110.7 Wiring Integrity.
393.	D	all of the above	210.52 Dwelling Unit Receptacle Outlets. (E) Outdoor Outlets. (1) One-Family and Two-Family Dwellings
394.	C	1000	410.140 General, (B) Dwelling Occupancies.
395.	C	evenly proportioned	210.11 Branch Circuits Required. (B) Load Evenly Proportioned Among Branch Circuits
396.	B	7	240.24 Location in or on Premises, (A) Accessibility.
397.	C	24, 40	310.120 Marking, B. (1) Surface Marking.
398.	D	equipment grounding conductor	215.6 Feeder Equipment Grounding Conductor.
399.	D	concrete tight	344.42 Couplings and Connectors. (A)
400.	C	100	215.3 Overcurrent Protection. Exception No. 1
401.	A	This is True	210.23 Permissible Loads. (A) 15- and 20- Ampere Branch Circuits.
402.	C	Overload	ARTICLE 100 Definitions, Overload
403.	B	Controller	ARTICLE 100 Definitions
404.	C	100, 40	220.82 Dwelling Unit. ((B) General Loads.
405.	A	permitted to be omitted	210.50 General. Receptacle outlets shall be installed as specified in 210.52 through 210.63. (B) Cord Connections.
406.	A	varyingly restricted	90.8 Wiring Planning. (B) Number of Circuits in Enclosures.

407.	C	not be used for purposes other than grounding	406.10 Grounding-Type Receptacles, Adapters, Cord Connectors, and Attachment Plugs, (C) Grounding Terminal Use.
408.	A	detached garage with electric power	210.52 Dwelling Unit Receptacle Outlets. (G) (1)) Garages, and Accessory Buildings.
409.	C	simultaneously	215.2 Minimum Rating and Size, (B) Feeders Over 600 Volts. (2) Feeders Supplying Transformers and Utilization Equipment.
410.	C	indicating	424.21 Switch and Circuit Breaker to Be Indicating.
411.	c	120	210.6 Branch-Circuit Voltage Limitations. (A) Occupancy Limitation.
412.	B	service conductors	230.7 Other Conductors in Raceway or Cable.
413.	C	8	250.53 Grounding Electrode System Installation, (G) Rod and Pipe Electrodes.
414.	A	maximum	210.19 Branch-Circuit Ratings Conductors Minimum Ampacity and Size. General A, 1(b)
415.	A	3	680.7 Cord-and-Plug-Connected Equipment, (A) Length.
416.	A	non-lighting	210.23 Permissible Loads. (D) Branch Circuits Larger Than 50 Amperes.
417.	C	7000, 7000 X 3.5 = 24,500	220.12: Table 220.12 General Lighting Loads by Occupancy
418.	B	1/4	312.3 Position in Wall.
419.	C	No solder shall be	410.121 Solder Prohibited.
420.	C	360	358.26 Bends - Number in One Run.
421.	E	Any of the above and more options are available	200.6 Means of Identifying Grounded Conductors.
422.	A	lighting outlet	210.70 Lighting Outlets Required. Lighting outlets shall be installed where specified in 210.70(A) (3) Storage or Equipment Spaces.
423.	C	1/2 inch	408.56: Table 408.56 Minimum Spacings Between Bare Metal Parts
424.	A	six	225.33 Maximum Number of Disconnects. (B) Single-Pole Units.

425.	B	2	220.12: Table 220.12 General Lighting Loads by Occupancy
426.	A	be listed for such use	110.14 Electrical Connections. (B) Splices.
427.	B	1/8	312.4 Repairing Noncombustible Surfaces. Noncombustible
428.	B	89	Annex B: Table B.310.15(B)(2)(3) Ampacities of Multiconductor Cables with Not More Than Three Insulated Conductors, Rated 0 Through 2000 Volts, in Free Air Based on Ambient Air Temperature of 40°C (104°F) (for Types TC, MC, MI, UF, and USE Cables
429.	B	1.5 ft.	210.52 Dwelling Unit Receptacle Outlets: A: (3) Floor Receptacles.
430.	B	concealed knob-and-tube wiring	314.3 Nonmetallic Boxes.
431.	E	All of the above	210.6 Branch-Circuit Voltage Limitations. (B) 120 Volts Between Conductors
432.	b	Circuit Breaker	ARTICLE 100 Definitions, Circuit Breaker.
433.	D	connected	250.112 Specific Equipment Fastened in Place (Fixed) or Connected by Permanent Wiring Methods. (M) Metal Well Casings.
434.	B	wall switch-controlled lighting	210.70 Lighting Outlets Required.(A) (1) Habitable Rooms.
435.	C	40	Chapter 9 Tables: Table 4 Dimensions and Percent Area of Conduit and Tubing (Areas of Conduit or Tubing for the Combinations of Wires Permitted in Table 1, Chapter 9)

Section 2 Timed Exams Answers

Timed Exam 1	1	C	9.5	Chapter 9 Tables, Table 2 Radius of Conduit and Tubing Bends
Timed Exam 1	2	C	538 sq ft	694.23 Turbine Shutdown. (A) Manual Shutdown.
Timed Exam 1	3	C	15	225.18 Clearance from Ground. (3)
Timed Exam 1	4	A	3, 8	410.10 Luminaries (Fixtures) in Specific Locations, (D) Bathtub and Shower Areas.
Timed Exam 1	5	C	3	ARTICLE 100 Definitions, Continuous Load
Timed Exam 1	6	B	35	Table C.1 Maximum Number of Conductors or FixtureWires in Electrical Metallic Tubing (EMT) (Based on Table 1, Chapter 9)
Timed Exam 1	7	C	as near as practicable to, and preferably in the same area as,	250.30 Grounding Separately Derived Alternating- Current Systems. (A) Grounded Systems. (4) Grounding Electrode.
Timed Exam 1	8	B	accessible	ARTICLE 100 Definitions, Accessible (as applied to equipment).
Timed Exam 1	9	D	12,200	Annex D, Example D3 Store Building
Timed Exam 1	10	B	40	374.5 Maximum Number of Conductors in Raceway
Timed Exam 1	11	C	1/2 inch	408.56: Table 408.56 Minimum Spacings Between Bare Metal Parts
Timed Exam 1	12	B	Multiple Fuse	ARTICLE 100 Definitions
Timed Exam 1	13	D	Fire-Resistive Cable System.	728.2 Definition. Fire-Resistive Cable System.
Timed Exam 1	14	D	general lighting	220.52 Small Appliance and Laundry Loads-Dwelling Unit. (B) Laundry Circuit Load.
Timed Exam 1	15	A	14	210.24: Table 210.24 Summary of Branch-Circuit Requirements
Timed Exam 1	16	E	all of the above	680.10 Underground Wiring

					Location.
	Timed Exam 1	17	B	Disconnecting Means	ARTICLE 100 Definitions
	Timed Exam 1	18	C	equipment bonding jumpers	250.98 Bonding Loosely Jointed Metal Raceways.
	Timed Exam 1	19	C	300	404.8 Accessibility and Grouping, (B) Voltage Between Adjacent Devices.
	Timed Exam 1	20	D	distinguishable by color, tagging, or other effective means from those that are part of the fire-rated design	300.11 Securing and Supporting, (A) Secured in Place (1) Fire-Rated Assemblies.
	Timed Exam 1	21	B	concealed knob-and-tube wiring	314.3 Nonmetallic Boxes.
	Timed Exam 1	22	C	permitted as the sole support	300.11 Securing and Supporting, (A) Secured in Place.
	Timed Exam 1	23	A	inadvertent	95.4 Continuity of Power.
	Timed Exam 1	24	D	A device intended for the protection of personnel that functions to deenergize a circuit or portion thereof within an established period of time when a current to ground exceeds the values established for a Class A device.	ARTICLE 100 Definitions; Ground-Fault Circuit Interrupter (GFCI).
	Timed Exam 1	25	B	42, six 7-A units 6x7=42	Annex D, Example D2(b) Optional Calculation for One-Family Dwelling, Air Conditioning Larger Than Heating
	Timed Exam 1	26	A	Location, Wet.	ARTICLE 100 Definitions, Location, Wet.
	Timed Exam 1	27	C	30, 90	110.26 Spaces, (2) Width of Working Space.
	Timed Exam 1	28	B	1/8	312.4 Repairing Noncombustible Surfaces. Noncombustible
	Timed Exam 1	29	A	dry walls	368.10 Uses Permitted, (C) (1) Walls.
	Timed Exam 1	30	C	Lead sheath	310.13: Table 310.104(A) Conductor Applications and Insulations Rated 600 Volts
	Timed Exam 1	31	A	tapped hole	314.40 Metal Boxes, Conduit Bodies, and Fittings, (D) Grounding Provisions.
	Timed Exam 1	32	D	four	220.53 Appliance Load Dwelling Unit(s).

Timed Exam 1	33	B	Intersystem Bonding Termination.	ARTICLE 100 - DEFINITIONS
Timed Exam 1	34	A	calculated	210.11 Branch Circuits Required. (A) Number of Branch Circuits.
Timed Exam 1	35	B	white	200.9 Means of Identification of Terminals.
Timed Exam 1	36	A	8, 3	300.14 Length of Free Conductors at Outlets, Junctions, and Switch Points.
Timed Exam 1	37	C	6, 8	230.202 Service-Entrance Conductors. (A) Conductor Size.
Timed Exam 1	38	E	All of the above	210.6 Branch-Circuit Voltage Limitations. (B) 120 Volts Between Conductors
Timed Exam 1	39	C	Overload	ARTICLE 100 Definitions, Overload
Timed Exam 1	40	C	15	Table C.1 Maximum Number of Conductors or FixtureWires in Electrical Metallic Tubing (EMT) (Based on Table 1, Chapter 9)
Timed Exam 1	41	A	6	300.5: Table 300.5 Minimum Cover Requirements, 0 to 1000 Volts, Nominal, Burial
Timed Exam 1	42	A	six	225.33 Maximum Number of Disconnects. (B) Single-Pole Units.
Timed Exam 1	43	A	Luminaire	ARTICLE 100 - DEFINITIONS
Timed Exam 1	44	B	ungrounded conductors	422.47 Water Heater Controls.
Timed Exam 1	45	C	Effective Ground-Fault Current Path.	250.4 General Requirements for Grounding and Bond- ing. (A) Grounded Systems. (5) Effective Ground-Fault Current Path.
Timed Exam 1	46	D	open or closed	225.38 Disconnect Construction. (D) Indicating.
Timed Exam 1	47	C	two or more 20-ampere	210.52 Dwelling Unit Receptacle Outlets (B) Small Appliances., (1) Receptacle Outlets Served.
Timed Exam 1	48	C	Busway	368.2 Definition.

Timed Exam 1	49	D	equipment grounding conductor	215.6 Feeder Equipment Grounding Conductor.
Timed Exam 1	50	A	Suitable	300.31 Covers Required.
Timed Exam 1	51	D	raceway	300.15 Boxes, Conduit Bodies, or Fittings (J) Luminaries (Fixtures).
Timed Exam 1	52	C	connected to earth	250.4 General Requirements for Grounding and Bonding. (A) Grounded Systems (1) Grounding of Electrical Equipment.
Timed Exam 1	53	B	1/4	312.3 Position in Wall.
Timed Exam 1	54	C	grouping all phase conductors together	300.20 Induced Currents in Ferrous Metal Enclosures or Ferrous Metal Raceways. (A) Conductors Grouped Together.
Timed Exam 1	55	B	wall switch-controlled lighting	210.70 Lighting Outlets Required.(A) (1) Habitable Rooms.
Timed Exam 1	56	B	sealed	225.27 Raceway Seal.
Timed Exam 1	57	A	non-lighting	210.23 Permissible Loads. (D) Branch Circuits Larger Than 50 Amperes.
Timed Exam 1	58	C	16	Annex C: Table C.1 Maximum Number of Conductors or FixtureWires in Electrical Metallic Tubing (EMT) (Based on Table 1, Chapter 9)
Timed Exam 1	59	A	This is true	250.118 Types of Equipment Grounding Conductors, (9)
Timed Exam 1	60	C	indicating	424.21 Switch and Circuit Breaker to Be Indicating.
Timed Exam 1	61	A	opposite	366.56 Splices and Taps, (B) Bare Conductors.
Timed Exam 1	62	D	Both A and C	210.8 Ground-Fault Circuit-Interrupter Protection for Personnel. (A) Dwelling Units.
Timed Exam 1	63	C	electrically operated by a readily accessible, remote-control device in a separate building or structure.	225.52 Disconnecting Means. (A) Location.
Timed Exam 1	64	C	Type NM cable	230.43 Wiring Methods for

				1000 Volts, Nominal, or Less.
Timed Exam 1	65	B	RNC	250.118 Types of Equipment Grounding Conductors.
Timed Exam 1	66	A	without the use of splice boxes	300.5 Underground Installations (E) Splices and Taps.
Timed Exam 1	67	D	Inverse Time	ARTICLE 100 Definitions
Timed Exam 1	68	B	reduced	358.24 Bends, How Made.
Timed Exam 1	69	C	grounded	404.2 Switch Connections, (B) Grounded Conductors.
Timed Exam 1	70	C	convection principles	110.13 Mounting and Cooling of Equipment. (B) Cooling.
Timed Exam 1	71	C	29,200 VA, 122 A	Annex D, Example D2(b) Optional Calculation for One-Family Dwelling, Air Conditioning Larger Than Heating
Timed Exam 1	72	C	40	Article 348 Flexible Metal Conduit (FMC) and Chapter 9, Table 1
Timed Exam 1	73	C	No solder shall be	410.121 Solder Prohibited.
Timed Exam 1	74	C	42	220.84: Table 220.84 Optional Calculations Demand Factors for Three or More Multifamily Dwelling Units
Timed Exam 1	75	C	100, 40	220.82 Dwelling Unit. ((B) General Loads.
Timed Exam 1	76	E	Any of the above and more options are available	200.6 Means of Identifying Grounded Conductors.
Timed Exam 1	77	A	Bonding Conductor or Jumper	ARTICLE 100 Definitions
Timed Exam 1	78	C	7000, 7000 X 3.5 = 24,500	220.12: Table 220.12 General Lighting Loads by Occupancy
Timed Exam 1	79	D	entrance	600.5 Branch Circuits, (A) Required Branch Circuit.
Timed Exam 1	80	E	all of the above	680.22 Area Lighting, Receptacles, and Equipment, (A) Receptacles, (2) Circulation and Sanitation System, Location.

Timed Exam 2	1	A	Three 15-A, 2-wire or two 20-A, 2-wire circuits	Annex D, Example D1(a) One-Family Dwelling
Timed Exam 2	2	C	115	Table 310.15(B)(16) Allowable Ampacities of Insulated Conductors
Timed Exam 2	3	A	100	310.15: Table 310.15(B)(6) Conductor Types and Sizes for 120/240-Volt, 3-Wire, Single-Phase Dwelling Services and Feeders. Conductor Types RHH, RHW, RHW-2, THHN, THHW, THW, THW-2, THWN, THWN-2, XHHW, XHHW-2, SE, USE, USE-2
Timed Exam 2	4	B	Remainder over 50,000	220.42: Table 220.42 Lighting Load Demand Factors
Timed Exam 2	5	B	6	680.22 Lighting, Receptacles, and Equipment, (A) Receptacles. (3) Other Receptacles, Location.
Timed Exam 2	6	B	final grade	230.24 Clearances. (B) Vertical Clearance for Overhead Service Conductors
Timed Exam 2	7	B	service conductors	230.7 Other Conductors in Raceway or Cable.
Timed Exam 2	8	A	ungrounded	210.10 Ungrounded Conductors Tapped from Grounded Systems.
Timed Exam 2	9	C	support	362.12 Uses Not Permitted.
Timed Exam 2	10	A	In storage battery rooms	348.12 Uses Not Permitted
Timed Exam 2	11	C	14.5 feet	ARTICLE 680, SWIMMING POOLS, FOUNTAINS, AND SIMILAR INSTALLATIONS, Table 680.8 (A) Overhead Conductor Clearances
Timed Exam 2	12	B	2	220.12: Table 220.12 General Lighting Loads by Occupancy
Timed Exam 2	13	C	2	250.50 Grounding Electrode System, (4) Ground Ring.
Timed Exam 2	14	A	grounded	225.38 Disconnect Construction.

				Disconnecting, (C) Disconnection of Grounded Conductor.
Timed Exam 2 -	15	C	18,600	Annex D, Example D1(a) One-Family Dwelling
Timed Exam 2 -	16	C	simultaneously	215.2 Minimum Rating and Size, (B) Feeders Over 600 Volts. (2) Feeders Supplying Transformers and Utilization Equipment.
Timed Exam 2 -	17	A	Ampacity	ARTICLE 100 Definitions
Timed Exam 2 -	18	B	first installed on-site	225.56 Inspections and Tests. (A) Pre-Energization and Operating Tests.
Timed Exam 2 -	19	B	4.5	Chapter 9 Tables, Table 2 Radius of Conduit and Tubing Bends
Timed Exam 2 -	20	D	24	300.5: Table 300.5 Minimum Cover Requirements, 0 to 1000 Volts, Nominal, Burial in Millimeters (Inches)
Timed Exam 2 -	21	C	nominal voltage	110.4 Voltages.
Timed Exam 2 -	22	A	varyingly restricted	90.8 Wiring Planning. (B) Number of Circuits in Enclosures.
Timed Exam 2 -	23	A	secured	312.5 Cabinets, Cutout Boxes, and Meter Socket Enclosures (C) Cables.
Timed Exam 2 -	24	B	15	406.3 Receptacle Rating and Type, (B) Rating.
Timed Exam 2 -	25	D	connected	250.112 Specific Equipment Fastened in Place (Fixed) or Connected by Permanent Wiring Methods. (M) Metal Well Casings.
Timed Exam 2 -	26	A	service-entrance	230.44 Cable Trays.
Timed Exam 2 -	27	A	5	680.22 Area Lighting, Receptacles, and Equipment. (C) Switching Devices.
Timed Exam 2 -	28	D	Neutral Point.	ARTICLE 100 - DEFINITIONS
Timed Exam 2 -	29	C	factory assembled in parallel	240.8 Fuses or Circuit Breakers in Parallel.
Timed Exam 2 -	30	A	Operation at a substantially constant load for an indefinitely	ARTICLE 100 Definitions: Duty, Continuous

			long time.	
Timed Exam 2	31	A	or project therefrom	312.3 Position in Wall.
Timed Exam 2	32	A	located or shielded	240.41 Arcing or Suddenly Moving Parts (A)
Timed Exam 2	33	B	the required number of poles	404.11 Circuit Breakers as Switches.
Timed Exam 2	34	A	3 ft	210.52 Dwelling Unit Receptacle Outlets. (D) Bathrooms.
Timed Exam 2	35	B	208	424.35 Marking of Heating Cables.
Timed Exam 2	36	C	6	300.14 Length of Free Conductors at Outlets, Junctions, and Switch Points.
Timed Exam 2	37	D	D	250.102 Bonding Conductors and Jumpers, (A) Material.
Timed Exam 2	38	A	water meters or filtering devices and similar equipment	250.53 Grounding Electrode System Installation, (D) Metal Underground Water Pipe.(1) Continuity.
Timed Exam 2	39	B	12	225.18 Clearance from Ground. (2)
Timed Exam 2	40	B	Capable of being operated without exposing the operator to contact with live parts.	ARTICLE 100 Definitions: Externally Operable
Timed Exam 2	41	D	all these outlets require GFI protection	210.8 Ground-Fault Circuit-Interrupter Protection for Personnel. (6) Kitchens
Timed Exam 2	42	C	corrosive	250.62 Grounding Electrode Conductor Material.
Timed Exam 2	43	C	1000	410.140 General, (B) Dwelling Occupancies.
Timed Exam 2	44	D	firestopped	300.21 Spread of Fire or Products of Combustion.
Timed Exam 2	45	A	This is True	210.23 Permissible Loads. (A) 15- and 20-Ampere Branch Circuits.
Timed Exam 2	46	A	9 sq. in.	680.26 Equipotential Bonding. (C) Pool Water.
Timed Exam 2	47	A	material of the conductor	110.14 Electrical Connections.
Timed Exam 2	48	A	no wiring systems of any type	300.22 Wiring in Ducts, Plenums, and Other Air-Handling Spaces, (A) Ducts for Dust, Loose Stock, or

				Vapor Removal.
Timed Exam 2 -	49	D	all of the above	210.52 Dwelling Unit Receptacle Outlets. (E) Outdoor Outlets. (1) One-Family and Two-Family Dwellings
Timed Exam 2 -	50	C	grounding electrode conductor	250.24 Grounding Service-Supplied Alternating Current Systems. (A) System Grounding Connections.
Timed Exam 2 -	51	D	circular mils	110.6 Conductor Sizes.
Timed Exam 2 -	52	C	Mechanical protection	225.20 Mechanical Protection of Conductors.
Timed Exam 2 -	53	B	panelboard	210.4 Multiwire Branch Circuits. (A) General.
Timed Exam 2 -	54	C	125	645.5 Supply Circuits and Interconnecting Cables, (A) Branch-Circuit Conductors.
Timed Exam 2 -	55	D	Practical Safeguarding.	ARTICLE 90 Introduction
Timed Exam 2 -	56	D	grounding electrode system	250.106 Lightning Protection Systems.
Timed Exam 2 -	57	B	6	250.53 Grounding Electrode System Installation, (B) Electrode Spacing.
Timed Exam 2 -	58	B	5	366.30 Securing and Supporting, (A) Sheet Metallic Auxiliary Gutters.
Timed Exam 2 -	59	B	conform with 490.24.	110.36 Circuit Conductors.
Timed Exam 2 -	60	D	145	Table 310.60(C)(69) Ampacities of Insulated Single Copper Conductor Isolated in Air Based
Timed Exam 2 -	61	A	CO/ALR	406.3 Receptacle Rating and Type, (C) Receptacles for Aluminum Conductors.
Timed Exam 2 -	62	D	grounded	230.22 Insulation or Covering. Exception:
Timed Exam 2 -	63	A	grounded	240.23 Change in Size of Grounded Conductor.
Timed Exam 2 -	64	D	wall switch controlled lighting outlet	210.70 Lighting Outlets Required. (A) (1) Habitable Rooms.
Timed Exam 2 -	65	C	10, 6	440.64 Supply Cords.
Timed Exam 2 -	66	A	3	230.9 Clearances on Buildings, (A) Clearances.

Timed Exam 2	67	D	the same	250.58 Common Grounding Electrode.
Timed Exam 2	68	D	Article 682	Table 250.3 Additional Grounding and Bonding Requirements, Article 682
Timed Exam 2	69	A	at the point where the conductors receive their supply	240.92 Location in Circuit. A.
Timed Exam 2	70	B	the ceiling assembly, including the ceiling support wires	300.11 Securing and Supporting, (A) Secured in Place (1) Fire-Rated Assemblies.
Timed Exam 2	71	C	Flexible Metal Conduit (FMC)	348.2 Definition.
Timed Exam 2	72	D	independently	240.22 Grounded Conductor.
Timed Exam 2	73	B	45	250.53 Grounding Electrode System Installation, (G) Rod and Pipe Electrodes.
Timed Exam 2	74	C	1.25, .0625	300.4 Protection Against Physical Damage. (A) Cables and Raceways Through Wood Members. (1) Bored Holes.
Timed Exam 2	75	C	12 in., 24 in.	210.52 Dwelling Unit Receptacle Outlets. (C) Countertops (1) Wall Counter Spaces.
Timed Exam 2	76	b	Circuit Breaker	ARTICLE 100 Definitions, Circuit Breaker.
Timed Exam 2	77	E	none of the above	410.151 Installation, (A) Lighting Track, (C) Locations Not Permitted.
Timed Exam 2	78	A	1 VA	220.12: Table 220.12 General Lighting Loads by Occupancy
Timed Exam 2	79	A	three, 24	310.15 Ampacities for Conductors Rated 0-2000 Volts. (A) General (3) Adjustment Factors.
Timed Exam 2	80	B	Bonding Jumper	ARTICLE 100 Definitions, Bonding Conductor or Jumper.
Timed Exam 3	1	B	2	210.52 Dwelling Unit Receptacle Outlets, (A) General Provisions. (2) Wall Space. (1)
Timed Exam 3	2	A	Where protection of the contained conductors is required from vapors, liquids, or solids	356.10 Uses Permitted. (2)

Timed Exam 3	3	A	white or gray	200.6 Means of Identifying Grounded Conductors. (C) Flexible Cords.
Timed Exam 3	4	D	Within a building	353.12 Uses Not Permitted.
Timed Exam 3	5	A	copper	110.5 Conductors.
Timed Exam 3	6	B	damp	406.9 Receptacles in Damp or Wet Locations. (A) Damp Locations.
Timed Exam 3	7	B	General-Use Switch	ARTICLE 100 Definitions
Timed Exam 3	8	C	24, 40	310.120 Marking, B. (1) Surface Marking.
Timed Exam 3	9	C	Orange	110.15 High-Leg Marking.
Timed Exam 3	10	B	double	314.16 Number of Conductors in Outlet, Device, and Junction Boxes, and Conduit Bodies. (4) Device or Equipment Fill.
Timed Exam 3	11	A	locked or sealed	230.93 Protection of Specific Circuits.
Timed Exam 3	12	B	The installation is in violation of Code requirements because the height requirement is not met., must be 17 feet minimum.	225.60 Clearances over Roadways, Walkways, Rail, Water, and Open Land, Table 225.61 Clearances over Buildings and Other Structures
Timed Exam 3	13	A	lighting outlet	210.70 Lighting Outlets Required. Lighting outlets shall be installed where specified in 210.70(A) (3) Storage or Equipment Spaces.
Timed Exam 3	14	B	ground faults	110.7 Wiring Integrity.
Timed Exam 3	15	C	Moisture- and heat-resistant thermoplastic	300.13: Table 310.104(A) Conductor Applications and Insulations Rated 600 Volts
Timed Exam 3	16	C	3	220.12: Table 220.12 General Lighting Loads by Occupancy
Timed Exam 3	17	C	Installations of communications equipment under the exclusive control of communications utilities located outdoors or in building spaces used exclusively	90.2 Scope (B) Not Covered. (4)

			for such installations	
Timed Exam 3	18	C	1 1/2	300.1 Scope, (C) Metric Designators and Trade Sizes.
Timed Exam 3	19	A	independent of the building	230.29 Supports over Buildings.
Timed Exam 3	20	D	range, counter-mounted cooking unit, or sink	210.52 Dwelling Unit Receptacle Outlets. (C) (4) Separate Spaces.
Timed Exam 3	21	D	all of the above	210.21: Table 210.21(B)(3) Receptacle Ratings for Various Size Circuits
Timed Exam 3	22	C	bathrooms	230.70 (2) Bathrooms.
Timed Exam 3	23	D	braces or guys	230.28 Service Masts as Supports. (A)
Timed Exam 3	24	D	15,510, Heat Pump and Supplementary Heat 240 V × 24 A = 5760 VA 15 kW Electric Heat, 5760 VA + (15,000 VA × 65%) = 5.76 kVA + 9.75 kVA = 15.51 kVA, if supplementary heat is not on at same time as heat pump, heat pump kVA need not be added to total.	Annex D, Example D2(c) Optional Calculation for One-Family Dwelling with Heat Pump (Single-Phase, 240/120-Volt Service) (see 220.82)
Timed Exam 3	25	D	Service Point	ARTICLE 100 Definitions
Timed Exam 3	26	D	suitable for use	225.22 Raceways on Exterior Surfaces of Buildings or Other Structures.
Timed Exam 3	27	D	A reliable conductor to ensure the required electrical conductivity between metal parts required to be electrically connected.	ARTICLE 100 Definitions : Bonding Conductor or Jumper.
Timed Exam 3	28	D	3, 5	342.30 (A) Securely Fastened. (1) & (2)

Timed Exam 3	29	C	heavy-duty type	210.21 Outlet Devices. Outlet devices shall have an ampere rating that is not less than the load to be served and shall comply with 210.21(A) and (B).
Timed Exam 3	30	B	dry location	680.23 Underwater Luminaries (Lighting Fixtures). (B) Wet-Niche Luminaires. (6) Servicing.
Timed Exam 3	31	C	the applied voltage	230.30 Insulation.
Timed Exam 3	32	A	either C or D	225.25 Location of Outdoor Lamps.
Timed Exam 3	33	A	10	225.18 Clearance from Ground. (1)
Timed Exam 3	34	B	18	250.64 Grounding Electrode Conductor Installation, (A) Aluminum or Copper-Clad Aluminum Conductors.
Timed Exam 3	35	C	Fixture Wiring	402.3: Table 402.3 Fixture Wires
Timed Exam 3	36	B	125 percent	B) Overcurrent Device Ratings.
Timed Exam 3	37	C	environment	300.6 Protection Against Corrosion and Deterioration.
Timed Exam 3	38	B	ungrounded	225.31 Disconnecting Means.
Timed Exam 3	39	C	360	358.26 Bends - Number in One Run.
Timed Exam 3	40	C	13	Table 314.16(A) Metal Boxes
Timed Exam 3	41	C	100 percent	(D) Selectivity.
Timed Exam 3	42	C	In a two story building	362.10 Uses Permitted
Timed Exam 3	43	B	Service Drop	ARTICLE 100 Definitions
Timed Exam 3	44	C	automotive vehicles other than mobile homes and recreational vehicles	90.2 Scope. B. Not Covered. (1)
Timed Exam 3	45	B	simultaneously disconnect	210.7 Branch Circuit Receptacle Requirements: Multiple Branch Circuits.
Timed Exam 3	46	C	Voltage (of a circuit).	ARTICLE 100 Definitions
Timed Exam 3	47	A	shall be permitted to be connected in parallel	310.10 (H) Conductors in Parallel. (1) General.
Timed Exam 3	48	C	3-wire feeders	215.4 (A) Feeders with

					Common Neutral Conductor.
Timed Exam 3		49	A	removed	250.96 Bonding Other Enclosures, (A) General.
Timed Exam 3		50	A	10,125 VA 40500x.25=10,125	220.12: Table 220.12 General Lighting Loads by Occupancy,
Timed Exam 3		51	A	voltage	110.9 Interrupting Rating.
Timed Exam 3		52	C	48, 60	680.9 Electric Pool Water Heaters.
Timed Exam 3		53	B	twice	220.55 Electric Ranges and Other Cooking
Timed Exam 3		54	C	shall be counted once	314.16 Number of Conductors in Outlet, Device, and Junction Boxes, and Conduit Bodies. (1) Conductor Fill
Timed Exam 3		55	D	either inside or outside	225.32 Location.
Timed Exam 3		56	B	1000	110.54 Bonding and Equipment Grounding Conductors. (A) Grounded and Bonded.
Timed Exam 3		57	D	any of the above	250.62 Grounding Electrode Conductor Material.
Timed Exam 3		58	C	High pressure sodium parking lot lighting	210.2 Table 210.2: Specific-Purpose Branch Circuits
Timed Exam 3		59	D	be guarded or isolated	240.41 Arcing or Suddenly Moving Parts (B)
Timed Exam 3		60	B	8.5 inches	Table 312.6(B) Minimum Wire-Bending Space at Terminals
Timed Exam 3		61	C	continuous	300.13 Mechanical and Electrical Continuity Conductors, (A) General.
Timed Exam 3		62	C	unless installed as part of a listed tub or shower assembly	404.4 Damp or Wet Locations (C) Switches in Tub or Shower Spaces
Timed Exam 3		63	C	Connector, Pressure (Solderless)	ARTICLE 100 Definitions, Connector, Pressure (Solderless)
Timed Exam 3		64	D	Luminaire	ARTICLE 100 Definitions, Luminaire
Timed Exam 3		65	B	17	402.5: Table 402.5 Allowable Ampacity for Fixture Wires
Timed Exam 3		66	A	Yes	Annex D, Example D4(a) Multifamily Dwelling

Timed Exam 3 -	67	D	35,000	328.10 Uses Permitted, Medium Voltage Cable,
Timed Exam 3 -	68	A	6	300.5: Table 300.5 Minimum Cover Requirements, 0 to 1000 Volts, Nominal, Burial
Timed Exam 3 -	69	B	Controller	ARTICLE 100 Definitions
Timed Exam 3 -	70	C	cord-and-plug-connected portable	210.19 Branch-Circuit Ratings Conductors Minimum Ampacity and Size. (2) Branch Circuits with More than One Receptacle.
Timed Exam 3 -	71	B	Cutout	ARTICLE 100 Definitions
Timed Exam 3 -	72	B	125	422.10 Branch-Circuit Rating, 422.10 Branch-Circuit Rating. (A) Individual Circuits.
Timed Exam 3 -	73	A	not less than 100% of	210.21 (1) Outlet Devices. Outlet devices shall have an ampere rating that is not less than the load to be served and shall comply with 210.21(A) and (B).
Timed Exam 3 -	74	D	Round boxes	314.2 Round Boxes.
Timed Exam 3 -	75	D	incandescent luminaries	404.14 Rating and Use of Snap Switches, (E) Dimmer Switches.
Timed Exam 3 -	76	B	lockable in accordance with 110.25.	225.52 Disconnecting Means. (C) Locking.
Timed Exam 3 -	77	A	copper, aluminum, or copper-clad aluminum	250.62 Grounding Electrode Conductor Material.
Timed Exam 3 -	78	B	removal of a trim or cover	240.83 Marking, (A) Durable and Visible.
Timed Exam 3 -	79	C	24 in. , 7 ft , 18 in.	760.24 Mechanical Execution of Work. (B) Circuit Integrity (CI) Cable.
Timed Exam 3 -	80	D	all of the above makes this statement true	225.52 Disconnecting Means (F) Identification.
Timed Exam 4 -	1	D	Bonding jumpers	250.92 Services,
Timed Exam 4 -	2	B	5 ft	Solar Photovoltaic (PV) Storage Batteries 690.71 Installation. (H) Disconnects and Overcurrent Protection.
Timed Exam 4 -	3	D	18, 36	422.16 Flexible Cords. (B)(2) Specific Appliances.

Timed Exam 4	4	C	the sum of the nameplate ratings of the transformers supplied when only transformers are supplied	215.2 Minimum Rating and Size, (B) (1) Feeders Supplying Transformers.
Timed Exam 4	5	D	exposed and concealed	358.10 Uses Permitted, (A) Exposed and Concealed.
Timed Exam 4	6	B	ungrounded conductors	215.7 Ungrounded Conductors Tapped from Grounded Systems.
Timed Exam 4	7	C	flexible cords, flexible cables, and fixture wires	240.4 Protection of Conductors.
Timed Exam 4	8	C	five	334.24 Bending Radius.
Timed Exam 4	9	C	combustible material	110.18 Arcing Parts.
Timed Exam 4	10	C	20.25	Table 314.16(A) Metal Boxes 1 for all grounding wires, 1 for fixture stud, 1 for all clamp inside of box = 9X2.25 = 20.25
Timed Exam 4	11	A	a lower	225.40 Access to Overcurrent Protective Devices.
Timed Exam 4	12	B	guards	110.27 Guarding of Live Parts. (B) Prevent Physical Damage.
Timed Exam 4	13	C	25 ft	625.16 Means of Coupling. (C) Overall Cord and Cable Length.
Timed Exam 4	14	C	conductors	300.3 Conductors., (B) Conductors of the Same Circuit.
Timed Exam 4	15	A	six, six	225.33 Maximum Number of Disconnects. (A) General.
Timed Exam 4	16	D	A and B	250.52 Grounding Electrodes, (B) Not Permitted for Use as Grounding Electrodes..
Timed Exam 4	17	C	insulated	250.119 Identification of Equipment Grounding Conductors.
Timed Exam 4	18	A	Yes	210.52: Figure 210.52(C)(1) Determination of Area Behind Sink or Range.
Timed Exam 4	19	D	twice	314.16 Number of Conductors in Outlet, Device, and Junction Boxes, and Conduit Bodies. (1) Conductor Fill

Timed Exam 4 -	20	A	25 ft	210.63 Heating, Air-Conditioning, and Refrigeration Equipment Outlet.
Timed Exam 4 -	21	D	nonabsorbent	230.27 Means of Attachment.
Timed Exam 4 -	22	D	closed in an approved manner	312.5 Cabinets, Cutout Boxes, and Meter Socket Enclosures. (A) Openings to Be Closed.
Timed Exam 4 -	23	D	Cable Management System (Electric Vehicle Supply Equipment)	625.2 Definitions.
Timed Exam 4 -	24	B	For direct burial where listed and marked for the purpose	350.10 (3) Uses Not Permited
Timed Exam 4 -	25	A	8, 3	225.19 Clearances from Buildings for Conductors of Not Over 1000 Volts, Nominal, (A) Above Roofs
Timed Exam 4 -	26	A	detached garage with electric power	210.52 Dwelling Unit Receptacle Outlets. (G) (1)) Garages, and Accessory Buildings.
Timed Exam 4 -	27	C	arc-fault circuit interrupter, combination-type,	210.12 Arc-Fault Circuit-Interrupter Protection. (A) Dwelling Units.
Timed Exam 4 -	28	D	none of the above	695.6 Power Wiring. (D) Pump Wiring.
Timed Exam 4 -	29	A	continuous	300.12 Mechanical Continuity Raceways and Cables.
Timed Exam 4 -	30	C	11.25	Table 310.15(B)(16) Allowable Ampacities of Insulated Conductors 25A X .45 (adjustment factor)= 11.25
Timed Exam 4 -	31	C	be permitted to serve as the disconnecting means	422.31 Disconnection of Permanently Connected Appliances, (A) Rated at Not Over 300 Volt-Amperes or 1/8Horsepower.
Timed Exam 4 -	32	D	126	Annex B: Table B.310.15(B)(2)(3) Ampacities of Multiconductor Cables with Not More Than Three Insulated Conductors, Rated 0 Through 2000 Volts, in Free Air Based on

					Ambient Air Temperature of 40°C (104°F) (for Types TC, MC, MI, UF, and USE Cables
Timed Exam 4	33	A	only one feeder or branch circuit	225.30 Number of Supplies.
Timed Exam 4	34	B	under a building	300.5 Underground Installations (C) Underground Cables Under Buildings.
Timed Exam 4	35	B	Permanent ladders or stairways	110.32 Work Space About Equipment. (B) Access.
Timed Exam 4	36	D	sealed or plugged at either or both ends	300.5 Underground Installations. (G) Raceway Seals.
Timed Exam 4	37	C	4500 VA	Annex D, Example D1(a) One-Family Dwelling
Timed Exam 4	38	A	a bushing or adapter	352.46 Bushings.
Timed Exam 4	39	E	All of the above	690.5 Ground-Fault Protection., (A) Ground-Fault Detection and Interruption.
Timed Exam 4	40	B	in parallel in each raceway or cable	250.122 Size of Equipment Grounding Conductors, (F) Conductors in Parallel.
Timed Exam 4	41	C	In commercial garages, other than for supplying ceiling outlets or extensions to the area below the floor but not above	372.4 Uses Not Permitted. (3)
Timed Exam 4	42	B	7	240.24 Location in or on Premises, (A) Accessibility.
Timed Exam 4	43	D	20, 2	250.52 Grounding Electrodes. (A) Electrodes Permitted for Grounding. (3) Concrete-Encased Electrode.
Timed Exam 4	44	C	Cabinet	ARTICLE 100 Definitions, Cabinet.
Timed Exam 4	45	C	the total ampere ratings of such units and not on the total watts of the lamps	220.18 Maximum Loads, (B) Inductive and LED Lighting Loads.
Timed Exam 4	46	C	An unintentional, electrically conducting connection between an ungrounded conductor of an electrical circuit and the normally non-current-carrying conductors, metallic enclosures, metallic raceways, metallic equipment, or	ARTICLE 100 Definitions: Ground Fault

			earth.	
Timed Exam 4	47	C	90	110.40 Temperature Limitations at Terminations.
Timed Exam 4	48	A	Calculated Service load 29,200 VA - Minimum Service Rating 122A	Annex D, Example D2(b) Optional Calculation for One-Family Dwelling, Air Conditioning Larger Than Heating
Timed Exam 4	49	A	angle	348.42 Couplings and Connectors.
Timed Exam 4	50	A	1	110.26 Spaces About Electrical Equipment, (C) Entrance to and Egress from Working Space., (1) Minimum Required.
Timed Exam 4	51	B	Plenum.	ARTICLE 100 Definitions
Timed Exam 4	52	A	be listed for such use	110.14 Electrical Connections. (B) Splices.
Timed Exam 4	53	C	At the time of installation, by a distinctive white or gray marking at its terminations. This marking shall encircle the conductor or insulation.	200.6 Means of Identifying Grounded Conductors. (B) Sizes Larger Than 6 AWG. (4)
Timed Exam 4	54	C	75	366.56 Splices and Taps, (A) Within Gutters.
Timed Exam 4	55	A	1/8, 125	220.18 Maximum Loads, (A) Motor-Operated and Combination Loads.
Timed Exam 4	56	B	crowding	312.7 Space in Enclosures.
Timed Exam 4	57	C	8	250.50 Grounding Electrode System, (5) Rod and Pipe Electrodes.
Timed Exam 4	58	A	8	368.12 Busways, Uses Not Permitted. (E) Working Platform.
Timed Exam 4	59	C	outlets not exceeding 240 volts	210.8 Ground-Fault Circuit-Interrupter Protection for Personnel. (C) Boat Hoists.
Timed Exam 4	60	D	A or B	250.119 Identification of Equipment Grounding Conductors. (C) Flexible Cord.

Timed Exam 4	61	C	Between a dwelling and separate garage unit	340.12 Uses Not Permitted.
Timed Exam 4	62	C	100	220.42: Table 220.42 Lighting Load Demand Factors
Timed Exam 4	63	B	12	Table 344.30(B)(2) Supports for Rigid Metal Conduit
Timed Exam 4	64	A	40	225.6 Conductor Size and Support, (B) Festoon Lighting.
Timed Exam 4	65	C	100	225.39 Rating of Disconnect. (C) One-Family Dwelling.
Timed Exam 4	66	C	Tap Conductors	240.2 Definitions.
Timed Exam 4	67	C	Interactive System	ARTICLE 100 Definitions, Interactive System
Timed Exam 4	68	A	bonding of electrical equipment	250.4 General Requirements for Grounding and Bond- ing. (A) Grounded Systems. (3) Bonding of Electrical Equipment.
Timed Exam 4	69	C	a single grounding electrode system.	250.53 Grounding Electrode System Installation, (B) Electrode Spacing.
Timed Exam 4	70	A	3	680.7 Cord-and-Plug-Connected Equipment, (A) Length.
Timed Exam 4	71	A	Bypass Isolation Switch	ARTICLE 100 Definitions
Timed Exam 4	72	B	10	230.26 Point of Attachment.
Timed Exam 4	73	E	all are acceptable	230.50 Protection Against Physical Damage. (B) All Other Service-Entrance Conductors.
Timed Exam 4	74	A	.412	Chapter 9 Tables: Table 5 Dimensions of Insulated Conductors and Fixture Wires
Timed Exam 4	75	C	.25, 2	250.50 Grounding Electrode System, (7) Plate Electrodes.
Timed Exam 4	76	D	continuity	250.12 Clean Surfaces.
Timed Exam 4	77	B	1.5 ft.	210.52 Dwelling Unit Receptacle Outlets: A: (3) Floor Receptacles.

Timed Exam 4	78	A	copper, aluminum, or copper-clad aluminum	250.118 Types of Equipment Grounding Conductors.
Timed Exam 4	79	B	untrained persons	90.1 Purpose.(C) Relation to Other International Standards.
Timed Exam 4	80	C	serve more than one kitchen	210.52 Dwelling Unit Receptacle Outlets. (3) Kitchen Receptacle Requirements.
Timed Exam 5	1	C	where exposed to direct sunlight	328.12 Uses Not Permitted.
Timed Exam 5	2	C	For system voltages of 1000 volts maximum	360.10 Uses Permitted. FMT
Timed Exam 5	3	B	electrical continuity	250.96 Bonding Other Enclosures, (A) General.
Timed Exam 5	4	A	Yes	Table 250.3 Additional Grounding and Bonding Requirements, Article 650
Timed Exam 5	5	D	30	250.53 Grounding Electrode System Installation, (G) Rod and Pipe Electrodes.
Timed Exam 5	6	A	bonded together	250.92 Services. (A) Bonding of Equipment for Services
Timed Exam 5	7	A	maximum	210.19 Branch-Circuit Ratings Conductors Minimum Ampacity and Size. General A, 1(b)
Timed Exam 5	8	D	permanent moisture level	250.53 Grounding Electrode System Installation, (A) Rod, Pipe, and Plate Electrodes. 1.
Timed Exam 5	9	C	45	310.15: Table 310.15(B)(2)(a) Adjustment Factors for More Than Three Current-Carrying Conductors in a Raceway or Cable
Timed Exam 5	10	C	covered	406.9 Receptacles in Damp or Wet Locations. (A) Damp Locations.
Timed Exam 5	11	C	30	250.53 Grounding Electrode System Installation, (H) Plate Electrode.
Timed Exam 5	12	D	concrete tight	344.42 Couplings and Connectors. (A)
Timed Exam 5	13	C	exothermic welding process	250.64 Grounding Electrode Conductor Installation, (C)

					Continuous.
Timed Exam 5	14	D	Either A or B		210.21: Table 210.21(B)(2) Maximum Cord-and-Plug-Connected Load to Receptacle
Timed Exam 5	15	B	2		220.12: Table 220.12 General Lighting Loads by Occupancy
Timed Exam 5	16	C	3 ft.		408.18 Clearances, (A) From Ceiling.
Timed Exam 5	17	D	2, 2		230.6 Conductors Considered Outside the Building.
Timed Exam 5	18	D	55		Table 310.15(B)(16) Allowable Ampacities of Insulated Conductors
Timed Exam 5	19	B	12, 4 1/2		350.30 Securing and Supporting. (A) Securely Fastened.
Timed Exam 5	20	D	125		210.19 Branch-Circuit Ratings Conductors Minimum Ampacity and Size. (a)
Timed Exam 5	21	B	B		210.23 Permissible Loads. (B) 30-Ampere Branch Circuits.
Timed Exam 5	22	D	vertical		240.33 Vertical Position.
Timed Exam 5	23	D	Equipment Bonding Jumper		ARTICLE 100 Definitions, Bonding Conductor or Jumper
Timed Exam 5	24	A	only one		230.2 Number of Services.
Timed Exam 5	25	D	compensate		300.7 Raceways Exposed to Different Temperatures, (B) Expansion Fittings.
Timed Exam 5	26	C	1500 volt-amperes		220.52 (A) Small Appliance and Laundry Loads Dwelling Unit.
Timed Exam 5	27	C	not be connected to the heater circuits		210.52 Dwelling Unit Receptacle Outlets.
Timed Exam 5	28	B	10		250.52 Grounding Electrode System, (A) Electrodes Permitted for Grounding, (1) Metal Underground Water Pipe.
Timed Exam 5	29	C	18		300.5: Table 300.5 Minimum Cover Requirements, 0 to 1000

				Volts, Nominal, Burial
Timed Exam 5	30	D	white or gray	200.10 Means of Identification of Terminals. (D) Screw Shell Devices with Leads.
Timed Exam 5	31	C	2520	Annex D, Example D4(a) Multifamily Dwelling
Timed Exam 5	32	B	5 ft	680.22 Lighting, Receptacles, and Equipment.(B) Luminaires, Lighting Outlets, and Ceiling-Suspended (Paddle) Fans.(6) Low-Voltage Luminaires.
Timed Exam 5	33	B	9000	Annex D, Example D3 Store Building
Timed Exam 5	34	C	8	210.24: Table 210.24 Summary of Branch-Circuit Requirements
Timed Exam 5	35	C	10	250.52 Grounding Electrodes, (A) Electrodes Permitted for Grounding.(2) Metal Frame of the Building or Structure. (1)
Timed Exam 5	36	B	5000	220.54 Electric Clothes Dryers Dwelling Unit(s).
Timed Exam 5	37	C	a separate equipment grounding conductor shall be installed in the conduit.	Article 358 Electrical Metallic Tubing: Type EMT, 356.60 Grounding.
Timed Exam 5	38	c	120	210.6 Branch-Circuit Voltage Limitations. (A) Occupancy Limitation.
Timed Exam 5	39	D	aluminum or copper-clad aluminum	250.64 Grounding Electrode Conductor Installation, (A) Aluminum or Copper-Clad Aluminum Conductors.
Timed Exam 5	40	A	This is True to the Code	404.9 Provisions for General-Use Snap Switches (B) Grounding.
Timed Exam 5	41	D	6 ft 7 in.	404.8 Accessibility and Grouping, (A) Location.
Timed Exam 5	42	D	55	220.54: Table 220.54 Demand Factors for Household Electric Clothes Dryers
Timed Exam 5	43	B	14	210.24: Table 210.24 Summary of Branch-Circuit Requirements
Timed Exam 5	44	C	100	215.3 Overcurrent

	-				Protection. Exception No. 1
Timed Exam 5	-	45	D	80	310.15: Table 310.15(B)(2)(a) Adjustment Factors for More Than Three Current-Carrying Conductors in a Raceway or Cable
Timed Exam 5	-	46	A	usually a resistor	250.36 High-Impedance Grounded Neutral Systems.
Timed Exam 5	-	47	D	1/2	314.24 Depth of Boxes. (A) Outlet Boxes Without Enclosed Devices or Utilization Equipment.
Timed Exam 5	-	48	C	shall not be	700.19 Multiwire Branch Circuits.
Timed Exam 5	-	49	D	simultaneously	225.52 Disconnecting Means.(B) Type.
Timed Exam 5	-	50	C	125	215.3 Overcurrent Protection.
Timed Exam 5	-	51	E	none of them	Article 750 Energy Management Systems, 750.30 Load Management, (A) Load Shedding Controls.
Timed Exam 5	-	52	B	100	215.2 Minimum Rating and Size, (A) Feeders Not More Than 600 Volts. Exception: No. 1
Timed Exam 5	-	53	E	any of these	250.110 Equipment Fastened in Place (Fixed) or Connected by Permanent Wiring Methods
Timed Exam 5	-	54	B	diagram showing feeder details	215.5 Diagrams of Feeders.
Timed Exam 5	-	55	A	2625, 10500 x .25 = 2625	220.12: Table 220.12 General Lighting Loads by Occupancy
Timed Exam 5	-	56	C	200	310.15: Table 310.15(B)(6) Conductor Types and Sizes for 120/240-Volt, 3-Wire, Single-Phase Dwelling Services and Feeders. Conductor Types RHH, RHW, RHW-2, THHN, THHW, THW, THW-2, THWN, THWN-2, XHHW, XHHW-2, SE, USE, USE-2
Timed Exam 5	-	57	B	55	400.5: Table 400.5(A)(2) Ampacity of Cable Types SC, SCE, SCT, PPE, G, G-

				GC, and W.
Timed Exam 5	58	A	visible break contacts	225.51 Isolating Switches.
Timed Exam 5	59	C	dedicated	110.26 Spaces About Electrical Equipment (E) Dedicated Equipment Space.
Timed Exam 5	60	A	1	220.14 Other Loads All Occupancies, (G) Show Windows.
Timed Exam 5	61	C	30	250.53 Grounding Electrode System Installation, (F) Ground Ring.
Timed Exam 5	62	B	8 AWG copper or 6 AWG	230.31 Size and Rating. (B) Minimum Size.
Timed Exam 5	63	C	4	342.20 Size Intermediate Metal Conduit: Type IMC (B) Maximum,
Timed Exam 5	64	D	screw shell	410.90 Screw-Shell Type.
Timed Exam 5	65	A	2	250.50 Grounding Electrode System, (A) Electrodes Permitted for Grounding, (7) Plate Electrodes.
Timed Exam 5	66	C	three	362.10 Uses Permitted.
Timed Exam 5	67	D	55	Table 310.15(B)(16) Allowable Ampacities of Insulated Conductors
Timed Exam 5	68	D	Thermoset	300.13: Table 310.104(A) Conductor Applications and Insulations Rated 600 Volts
Timed Exam 5	69	D	service conductors	215.2 Minimum Rating and Size, (A) (3) Ampacity Relative to Service Conductors
Timed Exam 5	70	B	50	110.27 Guarding of Live Parts. A.
Timed Exam 5	71	D	3 m (10 ft)	411.5 Specific Location Requirements. (B) Pools, Spas, Fountains, and Similar Locations
Timed Exam 5	72	A	staggered	225.24 Outdoor Lampholders.
Timed Exam 5	73	C	not be used for purposes other than grounding	406.10 Grounding-Type Receptacles, Adapters, Cord Connectors, and Attachment Plugs, (C) Grounding Terminal Use.

Timed Exam 5	74	B	12 linear ft	210.62 Show Windows.
Timed Exam 5	75	A	not be required	702.12 Outdoor Generator Sets. (B) Portable Generators 15 kW or Less
Timed Exam 5	76	B	concrete-encased	250.68 Grounding Electrode Conductor and Bonding Jumper Connection to Grounding Electrodes. (A) Accessibility. Exception No. 1
Timed Exam 5	77	C	40	Chapter 9 Tables: Table 4 Dimensions and Percent Area of Conduit and Tubing (Areas of Conduit or Tubing for the Combinations of Wires Permitted in Table 1, Chapter 9)
Timed Exam 5	78	A	6000	Annex D, Example D3 Store Building
Timed Exam 5	79	C	all the above	320.12 Uses Not Permitted.
Timed Exam 5	80	C	evenly proportioned	210.11 Branch Circuits Required. (B) Load Evenly Proportioned Among Branch Circuits